Advances in Intelligent Systems and Computing

Volume 829

Series editor

Janusz Kacprzyk, Polish Academy of Sciences, Warsaw, Poland
e-mail: kacprzyk@ibspan.waw.pl

The series "Advances in Intelligent Systems and Computing" contains publications on theory, applications, and design methods of Intelligent Systems and Intelligent Computing. Virtually all disciplines such as engineering, natural sciences, computer and information science, ICT, economics, business, e-commerce, environment, healthcare, life science are covered. The list of topics spans all the areas of modern intelligent systems and computing such as: computational intelligence, soft computing including neural networks, fuzzy systems, evolutionary computing and the fusion of these paradigms, social intelligence, ambient intelligence, computational neuroscience, artificial life, virtual worlds and society, cognitive science and systems, Perception and Vision, DNA and immune based systems, self-organizing and adaptive systems, e-Learning and teaching, human-centered and human-centric computing, recommender systems, intelligent control, robotics and mechatronics including human-machine teaming, knowledge-based paradigms, learning paradigms, machine ethics, intelligent data analysis, knowledge management, intelligent agents, intelligent decision making and support, intelligent network security, trust management, interactive entertainment, Web intelligence and multimedia.

The publications within "Advances in Intelligent Systems and Computing" are primarily proceedings of important conferences, symposia and congresses. They cover significant recent developments in the field, both of a foundational and applicable character. An important characteristic feature of the series is the short publication time and world-wide distribution. This permits a rapid and broad dissemination of research results.

More information about this series at http://www.springer.com/series/11156

Wilfried Lepuschitz · Munir Merdan
Gottfried Koppensteiner · Richard Balogh
David Obdržálek
Editors

Robotics in Education

Methods and Applications for Teaching and Learning

 Springer

Editors
Wilfried Lepuschitz
Practical Robotics Institute Austria (PRIA)
Vienna, Austria

Richard Balogh
Slovak University of Technology (STU)
Bratislava, Slovakia

Munir Merdan
Practical Robotics Institute Austria (PRIA)
Vienna, Austria

David Obdržálek
Charles University
Prague, Czech Republic

Gottfried Koppensteiner
Practical Robotics Institute Austria (PRIA)
Vienna, Austria

ISSN 2194-5357   ISSN 2194-5365 (electronic)
Advances in Intelligent Systems and Computing
ISBN 978-3-319-97084-4   ISBN 978-3-319-97085-1 (eBook)
https://doi.org/10.1007/978-3-319-97085-1

Library of Congress Control Number: 2018952057

This Springer imprint is published by the registered company Springer Nature Switzerland AG
The registered company address is: Gewerbestrasse 11, 6330 Cham, Switzerland

Preface

We are glad to present the proceedings of the 9th International Conference on Robotics in Education (RiE) held in Qawra, St. Paul's Bay, Malta, during April 18–20, 2018. RiE is organized every year with the goal to provide the opportunity for the presentation of relevant novel research and development in a strongly multidisciplinary context in the educational robotics domain.

Educational robotics is an innovative way of increasing the attractiveness of science education and scientific careers in the view of young people. Robotics represents a multidisciplinary and highly innovative domain encompassing physics, mathematics, informatics, and even industrial design as well as social sciences. As a multidisciplinary field, it promotes the development of systems thinking and problem solving. Due to various application areas, teamwork, creativity, and entrepreneurial skills are required for the design, programming, and innovative exploitation of robots and robotic services. The fascination for autonomous machines and the variety of fields and topics covered make robotics a powerful idea to engage with. Robotics confronts the learners with the areas of science, technology, engineering, arts, and mathematics (STEAM) through the design, creation, and programming of tangible artifacts for creating personally meaningful objects and addressing real-world societal needs. Thus, young girls and boys can easily connect robots to their personal interests and share their ideas. As a consequence, it is regarded as very beneficial if engineering schools and university program studies include the teaching of both theoretical and practical knowledge on robotics. In this context, current curricula need to be improved and new didactic approaches for an innovative education need to be developed for improving the STEAM skills among young people. Moreover, an exploration of the multidisciplinary potential of robotics toward an innovative learning approach is required for fostering the pupils' and students' creativity leading to collaborative entrepreneurial, industrial, and research careers.

In these proceedings, we present methods and applications for teaching and learning in the field of educational robotics. The book offers insights into the latest research, developments, and results regarding curricula, activities, and their evaluation. Moreover, the book introduces interesting programming approaches as well as new applications, technologies, systems, and components for educational robotics. The presented applications cover the whole educative range, from elementary school to high school, college, university, and beyond, for continuing education and possibly outreach and workforce development. In total, 45 papers were submitted and 27 papers are now part of these proceedings after a careful peer review process. We would like to express our thanks to all authors who submitted papers to RiE 2018, and our congratulations to those whose papers were accepted.

This publication would not have been possible without the support of the RiE International Program Committee and the conference Co-Chairs. We also wish to express our gratitude to the volunteer students and staff of the partner organizations, which significantly contributed to the success of the conference. All of them deserve many thanks for having helped to attain the goal of providing a balanced event with a high level of scientific exchange and a pleasant environment. RiE 2018 was supported by the eSkills Malta Foundation, for which we thankfully express our gratitude. We acknowledge the use of the EasyChair conference system for the paper submission and review process. We would also like to thank Dr. Thomas Ditzinger, Parvathidevi Krishnan, Saranya Kalidoss, and Springer for providing continuous assistance and advice whenever needed.

Organization

Committee

Co-chairpersons

Richard Balogh	Slovak University of Technology in Bratislava, Slovakia
Angele Giuliano	AcrossLimits, Malta
Wilfried Lepuschitz	Practical Robotics Institute Austria, Austria
David Obdržálek	Charles University, Czech Republic

International Program Committee

Dimitris Alimisis	EDUMOTIVA—European Lab for Educational Technology, UK
Julian Angel-Fernandez	Vienna University of Technology, Austria
Ansgar Bredenfeld	Dr. Bredenfeld UG, Germany
Jenny Carter	De Montfort University, Leicester, UK
Dave Catlin	Valiant Technology, UK
G. Barbara Demo	Dipartimento Informatica, Universita Torino, Italy
Jean-Daniel Dessimoz	University of Applied Sciences and Arts Western Switzerland, Switzerland
Nikleia Eteokleous	Dept. of Educational Sciences, Frederick University Cyprus, Cyprus
Hugo Ferreira	Instituto Superior de Engenharia do Porto, Portugal
Paolo Fiorini	University of Verona, Italy
Carina Girvan	Cardiff University, UK
José Gonçalves	Instituto Politécnico de Bragança, Portugal

Grzegorz Granosik	Lodz University of Technology, Poland
Marianthi Grizioti	University of Athens, Greece
Ivaylo Gueorguiev	European Software Institute—Center Eastern Europe, Bulgaria
Alexander Hofmann	University of Applied Sciences Technikum Wien, Austria
Martin Kandlhofer	Graz University of Technology, Austria
Boualem Kazed	University of Blida, Algeria
Gottfried Koppensteiner	Practical Robotics Institute Austria, Austria
Tomáš Krajník	University of Lincoln, UK
Miroslav Kulich	Czech Technical University in Prague, Czech Republic
Chronis Kynigos	University of Athens, Greece
Lara Lammer	Vienna University of Technology, Austria
Andrej Lúčny	Comenius University in Bratislava, Slovakia
Martin Mellado	Instituto ai2, Universitat Politècnica de València, Spain
Munir Merdan	Practical Robotics Institute Austria, Austria
Michele Moro	University of Padova, Italy
Margus Pedaste	University of Tartu, Estonia
Pavel Petrovič	Comenius University in Bratislava, Slovakia
Alfredo Pina	Public University of Navarra, Spain
João Machado Santos	University of Lincoln, UK
Fritz Schmöllebeck	University of Applied Sciences Technikum Wien, Austria
Dietmar Schreiner	Vienna University of Technology, Austria
Ugo Solitro	University of Verona, Italy
Roland Stelzer	INNOC—Austrian Society for Innovative Computer Sciences, Austria
Pavel Varbanov	European Software Institute—Center Eastern Europe, Bulgaria
Markus Vincze	Vienna University of Technology, Austria
Igor M. Verner	Technion—Israel Institute of Technology, Israel
Francis Wyffels	Ghent University, Belgium
Nikoleta Yiannoutsou	University of Athens, Greece

Local Conference Organization

Annalise Duca	AcrossLimits, Malta
Kerry Freeman	AcrossLimits, Malta
Munir Merdan	Practical Robotics Institute Austria, Austria

Sponsors

eSkills Malta Foundation

Contents

Comprehensive View on Educational Robotics

roboterfabrik: A Pilot to Link and Unify German Robotics Education to Match Industrial and Societal Demands

Sami Haddadin[1], Lars Johannsmeier[1], Johannes Schmid[2], Tobias Ende[2], Sven Parusel[2], Simon Haddadin[2], Moritz Schappler[1], Torsten Lilge[1], and Marvin Becker[1(✉)]

[1] Institute of Automatic Control,
Gottfried Wilhelm Leibniz Universität Hannover, Hanover, Germany
becker@irt.uni-hannover.de
[2] Franka Emika GmbH, Munich, Germany

Abstract. In this paper we introduce a novel robotics education concept entitled *roboterfabrik*. This approach is already implemented as a pilot project in the German educational system. Overall, we promote establishing the first generation of *robotic natives*. For this we need to provide both practical and theoretical experience in robotics to young people and give them access to state-of-the art, high performance yet affordable industrial robotic technology. Specifically, our approach systematically connects different existing school types, universities as well as companies. It comprises specialized lectures at the university, certified workshops and Robothons which are derived from the hackathon concept, and modified to the demand of roboticists.

1 Introduction and State of the Art

The significance and relevance of robotics beyond the professional community to laymen and even the general public has increased significantly over the last decade. After having enabled the third industrial revolution, robotics already arrived in our households. Lawn-mowers, cleaning robots, camera drones and robot toys have become a reality [28]. It is expected that over the next decade real service robots will enrich and facilitate our daily lives and change the way people interact with their environment. As of today collaborative robots are about to advance industrial processes in essentially all sectors [28].

This vast progress ultimately leads to the conclusion that the current generation of children is the first to grow up with real world robotics technology, making them *robotic natives* (short: *robonatives*). This is similar to the current *digital natives* generation that grew up at the advent of digital devices and today are thoroughly familiar with smartphones, IoT-devices and an omnipresent interconnectivity between people. We coined the expression *robotic natives* in 2015

© Springer Nature Switzerland AG 2019
W. Lepuschitz et al. (Eds.): RiE 2018, AISC 829, pp. 3–17, 2019.
https://doi.org/10.1007/978-3-319-97085-1_1

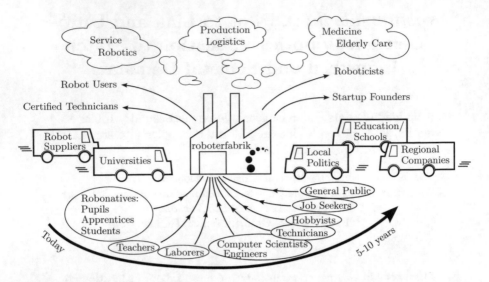

Fig. 1. Holistic concept for teaching robotics in the educational system. Different groups enter our knowledge factory *roboterfabrik*. Starting with teachers, pupils, apprentices and students in the education sector, it shall educate further societal groups in the future. Enablers such as (technical) universities, the industrial sector, local politics, the education system and robot manufacturers work closely together. This leads to qualified individuals creating benefits in different sectors by increased productivity as well as new business and technology ideas.

in the framework of our project *roboterfabrik*[1], which underlying concept and first results are elaborated in this paper. *roboterfabrik* aims for improving the robotics education of apprentices, high-school and university students starting in the region of Hanover, Germany as a pilot project [11,12]. Our mid-term vision is to purposefully and responsibly promote this new generation of *robonatives* with suitable educational concepts. The main goal is to enable them to use and further develop state-of-the-art robotic technology, create benefits for their own lives and careers and in turn help to shape our future society.

It is both necessary and advantageous for the general public as a whole to be familiarized with robotics. Industry as well as service providers, e.g. in healthcare, will inevitably utilize robots for many dull, dangerous and time consuming tasks. Hence, technicians, computer scientists and engineers with the knowledge to program and deploy robots at various technical levels will become a common job. Moreover, since robots will become part of our everyday lives, people of various backgrounds need to be accustomed to the devices.

---

[1] "roboterfabrik" is the literal German translation for "robot factory". More information can be found on the project website https://www.roboterfabrik.uni-hannover.de.

However, today's perception of robotics in society is well known to be rather complex. In surveys, people feel to a large extent neutral or positive when being questioned about service robots entering their homes [1,9,23]. At the same time, various gradations of fear are also reported [26]. More caution is found especially in the less technology-affine parts of society [5]. However, it is also known to be possible to reduce these effects by exposure to robots in the media [27] and even more by allowing experience with real robots [22]. Generally, the acceptance of robotics is not yet sufficiently studied due to biased studies [5] and unclear causalities. Furthermore, possible application areas and attitudes of general public towards robotics differ not only among countries [21] but also between group affiliations.

Generally, the robotics community has the obligation to actively inform and educate the general public [19] in order to correct expectations and avoid popular misunderstandings invoked by

1. the popular statement of robots being competitors to human workforce [8], instead of tools for humans to increase productivity, quality of life, as well as counteract the global resource distribution problem,
2. a false, often highly exaggerated expectation towards robotics induced by movies and science-fiction literature.

In this paper, we argue that these challenges should be tackled by

1. implementing a holistic approach for teaching robotics, ranging from basic education to further trainings and qualifications or workshops,
2. creating a way for the general public to encounter robots live while demonstrating possibilities and limitations.

We argue that such action would increase the motivation to use novel robotics technology and thus also societal acceptance. Consequentially, the demand for skilled workers and technical experts will increase even further, which is already intensified by the demographic change [16].

In this sense, our concept *roboterfabrik* starts at school level, where robotics may also increase the motivation to study STEM[2] subjects [4,10]. At university level, robotics is typically started as part of the graduate level curriculum of mechanical, electrical or computer engineering faculties. Recently, the field experiences increased popularity in student numbers and some universities now offer dedicated Bachelors or Masters programs [6]. From analyzing robotics education at university level [6], however, it becomes clear that laboratory experience with highly capable articulated manipulator systems is vital. However, despite vast progress the transformation of robots from basic positioning mechanisms to work and live companions has just begun and certainly not yet arrived at the education systems. In direct consequence, our education may profit vastly from the next generation of lightweight, safe robot systems, of which the Franka Emika *Panda* is the first one [7]. Therefore, it serves as the platform of choice in our Hanover pilot project. Besides its ability to safely and sensitively interact

---

[2] Science, technology, engineering and mathematics.

due to its soft-robotics paradigm as well as its unmatched affordability, its App-based programming interface allows the use of cutting edge robot technology in hands-on university and even high-school projects called "Robothon". Since the learning time for handling and programming the system is very short, the project tasks may even include rather complex human-robot interaction.

Contribution: In this paper we introduce and discuss the educational concept *roboterfabrik*, which

1. is a pilot implementation of a holistic robotics education in the German system
2. is sought to raise the acceptance and thus general understanding of robotics in society
3. provides a methodology to improve understanding for robotics and create a certain level of expertise in the general public

The *roboterfabrik* also aims to

- establish the term *robonatives*,
- introduce Robothons (= Robotics + Marathon) as a concrete robotics education tool,
- act as a platform to share new learnings and experiences with the robotics community.

The remainder of the paper is organized as follows: The *roboterfabrik* concept and its current structural elements are introduced in Sect. 2. The evaluation of completed Robothons and further development of future ones is described in Sect. 3. The pilot implementation of our concept in the German educational system is presented as a use-case in Sect. 4. Finally, Sect. 5 concludes the paper.

2   Concept

In this section, we describe the vision of the *roboterfabrik* pilot project, briefly review the robot platforms used for validation, and introduce the idea of certified workshops and robothons as a key to success.

2.1   Robotics in the Education System

Our approach brings robotics closer to the general public, see Fig. 1. It includes all interest groups, enablers and paths of possible outcome. Robotics is sought to be included in the curricula of apprentices, high-school and university students, and stimulate participation of teachers. Laborers, technicians and job-seekers improve their qualification with certified courses, while hobbyists and the general public may experience robotic technology at first hand. Apprentices, high-school and university students will naturally benefit from advanced robotics skills in their careers and will be considered the first generation *robonatives* that followed a systematic robotics education path, see Fig. 2. Note that we

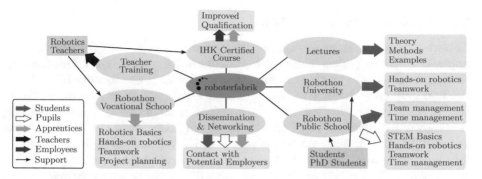

Fig. 2. The educational paths in our pilot project *roboterfabrik*. The light blue circles denote the available programs and grey boxes indicate the outcome for the different groups in the education system. The groups then connect the programs, e.g. teachers giving courses at vocational schools after they received a teacher training, or by students supervising pupils at school Robothons. (IHK = Industrie- und Handelskammer, see Sect. 4.2)

intentionally include vocational school teachers since they constitute a multiplier effect in particular in the German system. The overall knowledge increase will create significant benefits to existing and future workforce. Also, it will drive innovation and increase productivity in different sectors such as service robotics, manufacturing or healthcare.

To achieve this ambitious goal, several stakeholder of the German system need to collaborate closely:

- *Universities* develop the educational programs for theoretically oriented lectures and practical courses.
- *Robot technology suppliers* provide novel robot technology and give support.
- *Politics* provides funding for schools, brings the different parties together and initiates pilot projects.
- *Regional companies* connected in chambers of commerce and industry encourage their apprentices to get access to robot technology and sponsor the educational training of their staff.
- *Vocational schools* support teachers to acquire robotics know-how and transfer knowledge directly to the apprentices.

The developed connection of these stakeholders will be further elaborated below with the help of the specific pilot project in Hanover. This focuses on potential early adopters, as these have already some prior knowledge and show motivation in the expectation of better career opportunities in the future.

Next, we introduce the concrete implementation of our workshops including a detailed explanation of the Robothons.

2.2   Robotics Focus Workshops

Robotics focus workshops are our means of reaching out to high school students, apprentices or any person interested in robotics outside a university. Our goal for the participants is to be able to program a robot within one to three hours after the first encounter. In general, the workshops are organized in a modular structure that allows choosing the right topics according to the level of experience and demands of the different target groups. As already mentioned the Franka Emika *Panda* [7] system is the standard platform in the current pilot project. Therefore, our workshops focus on human-robot collaboration and assembly skills at the moment. The specific modules are grouped into the following categories:

- *General Theory*: Instructions for which no prior knowledge is required, such as a general introduction to robotics.
- *Technical Theory*: More advanced technical theory on robotics for which basic background knowledge is required or at least recommended, e.g. the high-level use of impedance control or safety methodologies.
- *Hands-on Experience*: Practical sessions with the robot platform including programming and teaching basic tasks. An important module of this group is the final project. This is tailored around the needs of the participants and fits the duration of the specific workshop.
- *Coding Modules with Robot Application*: Advanced programming with the robotic platform at hand. For this, a solid computer science background is obligatory, e.g. for programming new apps for the *Panda* system.
- *Exams*: Final modules in which the learning success of the workshop is tested and verified, including e.g. a final theoretical test for university students.
- *Trainer Modules*: Modules specialized to training the trainers who want to offer and teach their own workshops.
- *Customer Specific Modules*: Specialized modules for specific target groups, e.g. focusing on medical robot applications for caretakers.

The workshops usually take three to five days and the attendees receive a certificate after successful completion, see Sect. 4.2. Essentially, one could draw the analogy to a robot drivers license.

2.3   Robothons

Robothons are an essential part of the overall concept and central to the robotic focus workshops. They are designed to build upon a specific entry experience level. Hence, they can be adjusted to university students with strong focus on advanced robotics education, to high school students for more basic applications or to apprentices for specific practical applications.

The developed Robothons are similar to the well known concept of hackathons in which a technical problem is solved by a group of people mostly by programming in a specified amount of time. Unlike the interpretation of other authors [18,29], where a Robothon describes an event to build a working robot subject to a given specification, we aim at solving complete tasks and offer a well defined timely structure in small heterogeneous groups, e.g. by utilizing the *Panda* system, a 3D computer vision system and other supplementary equipment such as grippers, 3D printers, ...

Although the basic intention is always to give the students a hands-on experience in robotics, the specific topics of the Robothons may differ significantly. So far, two different types were implemented, the first focusing on human-robot collaboration and solving everyday tasks, while the other one relates to the application and development of machine learning algorithms for learning real-world assembly tasks. In the following, we refer to these specific Robothons as *collaborative Robothon* and *ML Robothon*, respectively.

Our concept stands out among others since instead of using rather toy-like systems with only marginal relation to real-world applications, we use cutting edge robot platforms that are currently introduced to real-world industry and research, i.e. the students practically experience highly relevant projects.

Typically, Robothons are integrated between a lecture and the accompanying exam. However, a Robothon may also be conceptualized as an independent module. The Robothon concept was first introduced at Technical University Munich (TUM) in 2011 as a block seminar in the lecture *Human Friendly Robotics*.

The general aim of a Robothon is to provide hands-on experience and inspire enthusiasm for novel technology that is about to hit industry. Additionally, the extensive need for teamwork promotes various soft skills. We further give students hands-on project planning experience by providing them with small project budgets to buy supplementary materials or supplies for their tasks.

The overall structure of the Robothons is outlined in the following.

Preparation Phase. Prior to beginning the actual work, the students assemble in groups of up to six people with heterogeneous backgrounds. From our experience, more students per group deteriorate effectiveness since a given task can only be divided into work packages to a certain degree. The Robothon supervisors are responsible for an equal distribution of skills among the different student groups such as programming, CAD design, or computer vision. For preparation the students may take any measures they deem necessary without already having access to a robot. This may include programming of software components that might be useful such as special computer vision algorithms, or the design and 3D-printing of gripper fingers that are specific to the respective task.

In addition, we provide the students with introductory crash-courses prior to the Robothon. For example, the CAD design and the design of gripper fingers or other special endeffector tools are particularly useful. The focus lies on enabling students to spot the important success factors in their assigned task, and how to match the robot capabilities with suitable finger and tool design. Specifically, depending on the required speed and space requirements of a task it might be more efficient to design a single endeffector or finger that can handle multiple situations instead of several ones that would need to be exchanged during the process.

Another introductory course elaborates on the used robot system itself, in our case Franka Emika's *Panda* [7]. In essence, this is a shortened version of the theoretical and practical units from the previously mentioned workshop we offer to high-school students and apprentices. Since students usually have a certain theoretical robotics background already, it is sufficient to teach them how to operate the robot. Particularly, this includes learning how to effectively teach it and how to use the web interface for developing novel solutions based on existing apps. Furthermore, the students learn how to program their own apps using a hierarchical hybrid statemachine-based programming. Although we provide a standard repertoire of various apps, individual tasks usually require the students to also write their own ones.

The third type of introductory course covers the implementation of computer vision algorithms and other functionalities via services written in C++. We provide a set of standard methods such as 3D-object detection and voice recognition. They enable the students to improve the prepared solution to match the needs of their own task. Nonetheless, if they desire to integrate and connect new algorithms and methods they are free to do so.

Eventually, in a fourth course the students are introduced to common machine learning tools and methods they may utilize in their Robothon. This includes popular and widely used software packages such as TensorFlow [24], Keras [2] and SciKit Learn [25].

Robothon Phase. The Robothon itself is divided into distinct phases to help the students to set goals and target advancements for the different days. This helps bringing certain structure to the student plans and encourages them to use a divide-and-conquer approach. Hence, the first day focuses on setting up the task i.e. placing all materials, involved devices and tools at the optimal location, experiencing the robot kinematics and thinking about possible solutions. In the end of the day the students present their plan, which comprises of specific work packages, milestones and assignments to group members. In the *collaborative Robothon*, we usually ask the students to solve their tasks first without the help of computer vision or other additional modalities, since this makes it easier for them to develop an initial working solution before integrating 3D vision solutions. In the *ML Robothon* a fully functional vision framework is provided as the focus of the course lies on the development of machine learning algorithms for force-sensitive assembly and insertions.

The second to fourth day contain mostly realization work, consisting of implementing the respective solution, writing skills and software if needed and improving working processes. The major milestone is to have a running application by the end of day four.

The last day is intended for optimization and implementation of additional elements that are not critical for the solution itself. Thereafter, the results are presented after a strict deadline. In addition, the developed approach together with a summary video are shown in a colloquium after the Robothon.

2.4   Reference Education Platform

Only with the availability of the intuitive, safe and affordable robot system Franka Emika *Panda* [7] it was made possible to initiate the *roboterfabrik* concept. This robot constitutes the core component and needs to be usable even by high-school students as young as fourteen years old, yet at the same time be capable of industrial grade automation. Moreover, for more advanced students who desire to graduate in robotics or some closely related field, they have to provide the necessary low-level interfaces to allow research activities and adding new features via advanced programming.

3D Perception technology such as Microsoft Kinect2 [20] and Intel RealSense SR300 [13] are also made available. The former is used as a static camera while the SR300 is mounted on the robot endeffector.

So far our setup is rather unique since there has not yet been a robot system that is both affordable and technologically relevant such as *Panda*. Most educational approaches in the robotics community make use of small mobile robots since these are usually much cheaper than complex manipulator systems, let alone soft-robotics enabled ones [6]. This of course results in major restrictions in terms of learning. Typically, the focus lies on navigation or low-level programming [14]. In contrary, we are able to offer for the first time an educational program including interaction, manipulation and learning for cutting edge soft-robots.

3   Evaluation

So far we were able to reach about 175 university students mostly from mechatronics, electrical engineering and computer science masters programs, as well as almost 100 students which were either in high school grade seven to twelve or first-year apprentices from vocational schools. The addressed topics span from human-robot collaboration in manufacturing to household robotics for the elderly. The number of Robothon participants has increased significantly over time. Impressions[3] from previous events are shown in Fig. 3.

Next, gained insights and possible evaluation metrics for future events are discussed.

3.1   Lessons Learned

The increasing number of completed Robothons provides us with the possibility to critically review the success of the events and use according feedback to continuously improve them. Lessons learned so far can be summarized as follows.

---

[3] Further impressions of various events and Robothons can be found at https://www.roboterfabrik.uni-hannover.de.

Fig. 3. Impressions from the opening of the *roboterfabrik* with the Minister President of Lower Saxony (left), previous *collaborative Robothons* (middle) and *ML Robothons* (right).

- A comparison of earlier Robothons without preparatory phase to the more recent ones clearly shows that students, who were introduced to the specific tools of the Robothons use more structured and well-thought-out approaches. This results in better planning at the beginning and significantly improved final results. Moreover, the comparison indicates that forming student groups should be done before the preparation phase. The safety instructions were done during the preparation phase, so the students learn early on how to safely use and interact with a robot.
- A planning phase in which the students plan their entire approach, divide their project into work packages, and distribute these among team members according to their particular skills and interests has turned out to be very important. Furthermore, due to the constrained available time frame for solving the problem at hand, an effective time management is required.
- It is necessary to provide selected software packages with high quality documentation to allow the groups a qualified decision process which packages to select for their respective tasks. Otherwise, the amount of available packages (especially the ones available in the popular ROS framework) constitutes a major hurdle. To use standardized and well-documented packages also allows the groups to work independently and decreases supervision efforts.
- It turned out to be more efficient to assign dedicated supervisors as specialists e.g. in 3D printing, computer vision or robot app development. Supervisors directly assigned to a group could hardly be trained in all hardware and software components. Thus, one or two specialists are required to provide assistance when specific questions or unforeseen problems arise.

3.2 Possible Evaluation Methods and Metrics

Among the possible *methods* for evaluating Robothons are anonymous feedback sheets as is common for all lectures at the *Gottfried Wilhelm Leibniz University Hanover*. Furthermore, structured interviews with the participants, as e.g. the ones performed in [17] to assess positive and negative influencing factors for the FH Salzburg Robothon, are used. The advantages of questionnaires are anonymity of the participants and their convenience and simplicity. Interviews on the other hand, typically provide more informations and allow to reassure given answers. However, they may be biased with respect to positive results, if the Robothon is part of graded lectures.

We used the questionnaire-type evaluation to inquire about equipment, safety measures, tasks, supervision, working atmosphere in the groups and the organization. The answers are linguistic and have to be transferred into a *metric representation* such as "good/average/bad" or "number of positive or negative occurrences of aspects" for statistical analysis. This, however is ongoing process and results will be published at a later stage.

4 Further Roll-Out

The roll-out of the *roboterfabrik* concept presented in Sect. 2 has started in the German pilot region Hanover. On a mid- to long-term perspective the education concept and its focus on robotics is sought to be transferred to the German dual education system (see Sect. 4.1) with a certification by the associated authorities (see Sect. 4.2). The planned transformation of the pilot project into a systematic roll-out is shortly described in Sect. 4.3.

4.1 Integration into the German Dual Education System

The dual education system in German-speaking countries is unique. After attending public school until lower secondary level for ten years, significant numbers of students become apprentices in companies, learn practical skills necessary for future jobs and attend vocational schools at the same time to learn corresponding background theory. The apprenticeship ends after two to three years with final examinations, including theoretical exams and practical works. This system has led to high employment rates of younger people and a good body of qualified workers for German industry. As already mentioned, our goal is to integrate the *roboterfabrik* concept into the German dual education system, see Fig. 4. The courses are designed for skilled workers as industrial trainings. By providing them also to apprentices, we improve their qualification and offer them a low-threshold opportunity to become familiar with the technology. The young skilled workers then become *robonatives* by our definition and bring cutting edge robotics know-how to small and medium enterprises (SMEs), which in contrast to large companies have not yet profited at large from automation [15].

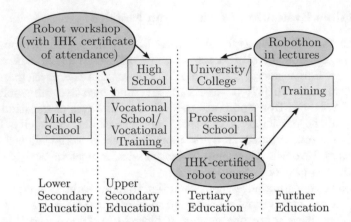

Fig. 4. Excerpt of the German education system with focus on institutions relevant for technical professions. Our approach to deliver robotics education to all levels of the system is accomplished by the robotics focus workshops (Sect. 2.2), Robothons (Sect. 2.3) and certified courses (Sect. 4.2). Solid arrow-lines point to the main targets.

4.2   Certification

The chambers of commerce and industry ("Industrie- und Handelskammer", "IHK") are regionally organized associations of businesses, which manage major parts of the dual education system in Germany. Additionally, they offer certified training sessions that are highly valued and accepted by companies for on-the-job training of their employees [3]. A certificate course may consist of 50 to 400 (cumulated) hours of training with defined measures of learning success e.g. by tests or work samples. The first set of results from the Robothons was integrated into new certificate courses on how to use collaborative robots. An "operator certificate" course is currently developed in close cooperation with the vocational school BBS Neustadt and Franka Emika GmbH. The course is created particularly for apprentices and skilled workers, empowering the participants to set up a robot for envisaged automation tasks. This includes creating a concept for solving the task, selecting and collocating robot apps, installing a test demonstrator, learning to make the application robust and integrating the setup into a factory ecosystem with particular focus on safety and efficiency. The course will be offered in the near future at the BBS Neustadt. Teachers of the vocational school already took training courses from Franka Emika GmbH in order to be able to safely handle the robot and teach relevant skills to their students. Parts of the certificate content will be taught in the regular curriculum, while interested and motivated students have the opportunity to take further modules in order to obtain the full certificate.

Meanwhile, we are working on an "expert certificate"-course for automation engineers and programmers, which focuses on the creation of new programs and robot applications. This includes programming the robot not only via teaching, but at code level. Participants are then able to include computer vision pipelines and create state machines for advanced apps. The content of this course is roughly equivalent to the student Robothons. Students who participated in the Robothons will also receive the certificate to increase the attractiveness of the model.

To create a certificate from the Robothon concept learning targets, contents, and materials have to be reviewed, prerequisites, tests, and evaluation of the participants properly defined. Finally, course evaluation and systematic feedback are intended for quality assurance.

4.3 From Pilot Project to General Concept

The systematic integration of robotics into the university, schools as well as apprentice education (see Fig. 4) is brought together in the project *roboterfabrik*. The first pilot in Hanover developed concepts and learning materials, and collected feedback for the improvement of the process. Meanwhile, the certificate courses are built up in the regional vocational schools with the help of the chamber of commerce and industry. We also work on distributing the idea and structure of the *roboterfabrik* nation-wide, starting in the state of Lower Saxony.

5 Conclusion

Our educational concept *roboterfabrik* is regarded as a significant step towards integrating robotics into our educational system. It already led to considerable interest among universities, schools, companies and politics alike. After the very successful pilot project in Hanover, we plan to transfer *roboterfabrik* to other locations. Based on the feedback and evaluations of previous events, we continuously extend and enhance the workshops and Robothons. New workshops will be tailored in the future around specific needs of important target groups such as e.g. logistics or healthcare. At some point this could enable us to make the *roboterfabrik* concept accessible even to the general public.

Acknowledgment. We would like to sincerely thank the Region Hannover for their generous funding of the project *roboterfabrik*. We would also like to give special thanks to Reinhard Biederbeck from Region Hannover, Torsten Temmeyer from IHK Hannover, the teachers from the regional vocational schools BBS ME and BBS Neustadt for their support in creating the certification of the workshops. Special thanks go to Ina May from Roberta RegioZentrum Hannover and the participating pupils and students for active engagement and continuous valuable feedback.

References

1. Arras, K.O., Cerqui, D.: Do we want to share our lives and bodies with robots? a 2000 people survey. Technical report (2005)
2. Chollet, F., et al.: Keras (2015). https://github.com/fchollet/keras. Accessed 15 Sept 2017

3. Deutscher Industrie- und Handelskammertag: Education and training (2017). https://www.dihk.de/en/segments/training/education-training. Accessed 15 Sept 2017
4. Eguchi, A.: Robotics as a learning tool for educational transformation. In: Proceeding of 4th International Workshop Teaching Robotics, Teaching with Robotics and 5th International Conference Robotics in Education, Padova, Italy, pp. 27–34 (2014)
5. Enz, S., Diruf, M., Spielhagen, C., Zoll, C., Vargas, P.A.: The social role of robots in the future—explorative measurement of hopes and fears. Int. J. Soc. Robot. **3**(3), 263 (2011)
6. Esposito, J.: The state of robotics education: proposed goals for positively transforming robotics education at postsecondary institutions. IEEE Robot. Autom. Mag. **24**, 157–164 (2017)
7. Franka Emika GmbH: FRANKA EMIKA (2017). https://www.franka.de/. Accessed 21 Aug 2017
8. Frey, C.B., Osborne, M.A.: The future of employment: how susceptible are jobs to computerisation? Technol. Forecast. Soc. Change **114**, 254–280 (2017)
9. German Federal Ministry of Education and Research: Service-Roboter statt Pflegeheim (results from representative phone interviews) (2016). https://www.bmbf.de/de/service-roboter-statt-pflegeheim-2727.html. Accessed 14 Sept 2017
10. Guizzo, E.: How to get students excited about engineering? Bring in the robots (review about a summer camp at GRASP lab) (2007). http://spectrum.ieee.org/automaton/robotics/robotics-software/how_to_get_students_excited_ab. Accessed 29 Aug 2017
11. Haase, B.: Hannover soll führender Robotikstandort werden (2015). http://www.haz.de/Hannover/Aus-der-Stadt/Uebersicht/Hannover-soll-fuehrender-Robotikstandort-werden. Accessed 29 Aug 2017
12. Haddadin, S., Johannsmeier, L., Becker, M., Schappler, M., Lilge, T., Haddadin, S., Schmid, J., Ende, T., Parusel, S.: Roboterfabrik: a pilot to link and unify german robotics education to match industrial and societal demands. In: Companion of the 2018 ACM/IEEE International Conference on Human-Robot Interaction, pp. 375–375. ACM (2018)
13. Intel: Intel RealSense SR300 (2016). https://software.intel.com/en-us/realsense/sr300. Accessed 29 Aug 2017
14. Jambor, T.N.: Techcolleges: learn to teach using robots. In: International Conference on Robotics and Education RiE 2017, pp. 3–14. Springer (2017)
15. Jäger, A., Moll, C., Som, O., Zanker, C.: Analysis of the impact of robotic systems on employment in the european union. Technical report, Fraunhofer Institute for Systems and Innovation Research ISI, Karlsruhe (2015)
16. Kopf, H., Müller, S., Rüede, D., Lurtz, K., Russo, P.: Einführung: made in Germany? Fachkräftemangel gefährdet den Wirtschaftsstandort Deutschland. In: Soziale Innovationen in Deutschland, pp. 61–63. Springer (2015)
17. Kriegseisen-Peruzzi, M.: Ergotherapeutisch-handlungswissenschaftliche Zugänge als Ressource in der Entwicklung neuer Technologien: Begleitstudie zum ersten Robothon an der FH Salzburg. Fachtagung Ergotherapie Austria (2016). https://www.fhg-tirol.ac.at/data.cfm?vpath=pdf-downloads/robothon-handout_-ftea_2016_akpdf. Accessed 29 Aug 2017
18. Merz, R.: Robothon Fachhochschule Salzburg (2015). http://its.fh-salzburg.ac.at/events/robothon/ueber-robothon/. Accessed 21 Aug 2017
19. Meyer, S.: Einsatz von Robotik in der Pflege - was zeichnet sich ab? In: Der Demographiekongress (2017)

20. Microsoft: Kinect V2 (2014). http://www.xbox.com/en-US/xbox-one/accessories/kinect. Visited on 29th of August 2017
21. Nomura, T., Suzuki, T., Kanda, T., Han, J., Shin, N., Burke, J., Kato, K.: What people assume about humanoid and animal-type robots: cross-cultural analysis between Japan, Korea, and the United States. Int. J. Humanoid Robot. **5**(01), 25–46 (2008)
22. Nomura, T., Suzuki, T., Kanda, T., Yamada, S., Kato, K.: Attitudes toward robots and factors influencing them. In: New Frontiers in Human-Robot Interaction, pp. 73–88 (2011)
23. Nomura, T., Tasaki, T., Kanda, T., Shiomi, M., Ishiguro, H., Hagita, N.: Questionnaire-based research on opinions of visitors for communication robots at an exhibition in Japan. Lect. Notes in Comput. Sci. **3585**, 685 (2005)
24. Open Source: TensorFlow: Large-scale machine learning on heterogeneous systems (2017). https://www.tensorflow.org/. Accessed 30 Aug 2017
25. Pedregosa, F., Varoquaux, G., Gramfort, A., Michel, V., Thirion, B., Grisel, O., Blondel, M., Prettenhofer, P., Weiss, R., Dubourg, V.: Scikit-learn: machine learning in python. J. Mach. Learn. Res. **12**, 2825–2830 (2011)
26. Ray, C., Mondada, F., Siegwart, R.: What do people expect from robots? In: IEEE/RSJ International Conference on Intelligent Robots and Systems, IROS 2008, pp. 3816–3821. IEEE (2008)
27. Riek, L.D., Adams, A., Robinson, P.: Exposure to cinematic depictions of robots and attitudes towards them. In: Proceedings of International Conference on Human-Robot Interaction, Workshop on Expectations and Intuitive Human-Robot Interaction (2011)
28. Sahi, M.K., Kaul, A.: Consumer robotics - household robots, vacuum robots, lawn mowing robots, pool cleaning robots, personal assistant robots, and toy and educational robots: Global market analysis and forecasts (2017)
29. Suzuki, K., Zhu, X.: Regions bustle with workshops, courses, robothon, and a society inauguration [chapter news]. IEEE Robot. Autom. Mag. **23**(4), 193–198 (2016)

The Robotics Concept Inventory

Reinhard Gerndt[1(✉)] and Jens Lüssem[2]

[1] Ostfalia University, Wolfenbuettel, Germany
r.gerndt@ostfalia.de
[2] University of Applied Sciences Kiel, Kiel, Germany
jens.luessem@fh-kiel.de

Abstract. Robotics became one of the main subjects in modern STEM (Science, Technology, Engineering and Mathematics) education and training. Many teaching approaches like lecture, hands-on exercises, project and problem-based and competition-driven courses compete for best learning outcome [1–6]. However, the community lacks a tool to assess the level of robotics-specific competencies of students for a better-targeted teaching and training and especially for a quantitative standardized assessment of the specific learning and teaching approaches. Concept inventories are such a tool and used in many fields already [7]. They allow assessment of students' understanding of the most central concepts of a domain. When students undergo the test at the beginning and the end of a course or training, the relative performance gives a numerical value for the teaching and learning gain and answers to distinct questions may highlight individual problems. With this paper we present the first version of a complete Robotics concept inventory.

Keywords: Robotics education · Concept inventory · Teaching assessment
Learning assessment

1 Introduction

Reasons for a normative assessment of learning and teaching are manyfold. Many different pedagogical approaches in teaching compete for the best learning results. Students with different backgrounds may require different approaches. As a consequence of the Bologna process and high expectations on student mobility, quality assurance is a crucial objective for many universities [4, 8–10]. Educational service providers for vocational training need to show the effectiveness of their services.

Currently, assessment and quality assurance in these contexts often is carried out by reviews and audits that often is biased by personal experience - internal or external - auditors or members of review committee. Moreover, we need a more agile reaction to new requirements by academia and industry.

Formative and lightweight methods, like concept inventories appear to be a better-suited tool for an assessment as could be shown with many existing concept inventories. However, what makes robotics different to the other, more pure, STEM subjects is the highly trans-disciplinary nature of robotics.

The remaining part of this paper is organized as follows: After this introduction we will shortly repeat the basics of concept inventories. Then we present a list of concepts

© Springer Nature Switzerland AG 2019
W. Lepuschitz et al. (Eds.): RiE 2018, AISC 829, pp. 18–27, 2019.
https://doi.org/10.1007/978-3-319-97085-1_2

that a central to the field of robotics and relate those to underlying categories. Eventually, we present the details of the robotics CI development process and sketch some CI test questions prior to a short summary. The authors will make available the full CI test with the complete list of questions upon request and evidence of teaching background, for reasons sketched in the next chapter.

2 Concept Inventories Revisited

In this section we will revisit the basics of concept inventories. For this, we will have a look into one of the most often referred to Signals and Systems concept inventories, which typically the first exposure of students to the fields of robotics or signal processing or transmission. The respective work started in the year 2000 and has initially been published by Wage et al. in her article 'The Signals and Systems Concept Inventory' in 2005 [11].

The first phase focused on collecting central concepts and designing the test. In a second alpha-test phase the CI tests have been verified to possibly revise the exam questions and provide an initial calibration of numerical results. The signals and systems concept inventory (SSCI) actually consists of a continuous time (CT) and a discrete time (DT) inventory. Figure 1 lists 6 categories for the DT-SSCI and 25 concepts, which are represented individually by 25 multiple-choice questions.

The categories shall cover the entire subject, however, the de-facto standard of 25 questions, stemming from the maximum duration of the test, limits the number of categories and concepts. The DT-SSCI concepts are related to mathematics (Math), linearity and time-invariance (LTI), sampling (Sampling), transformation (Trans), filtering (Filt) and convolution (Conv). Some concepts are related to more than one category. The one-out-of-four questions are designed to elicit if students actually understood the underlying concept. Figure 2 shows as example question 1 of the CT-SSCI.

Tests are graded on a scale between 0 (worst) and 100 (best), typically with 4 points for every correct answer and 0 points for a wrong or missing answer. From pre- and post-teaching test results, i.e. from the results of the test at the beginning of the course and the same (retaken) test at the end of the course, for every student an individual gain is calculated by the following formula:

$$\text{gain} = \frac{\text{post-pre}}{\text{100-pre}} \tag{1}$$

For an overall assessment, the grades are aggregated. Figure 3 shows an aggregated representation of different course types.

For more details on concept inventories please refer to some of the early work like [7, 11, 12].

#	Category	Concept(s)
1	Math	Time/frequency: select the plot of the sinusoid with the highest frequency
2	Math	Time-reversal: given a plot of $p[n]$, recognize the plot of $p[-n]$
3	Math	Time-shift: given a plot of $p[n]$, recognize the plot of $p[n-1]$
4	Math	Basic signals: recognize a plot of $u[n] - u[n-2]$
5	Math	Periodicity of DT sinusoids: given a plot of $\cos(\omega_0 n)$, recognize plot of $\cos((\omega_0 + 2\pi)n)$
6	LTI	Time invariance: given an input/output pair for an LTI system, recognize the output when the input is a shifted version of the original input
7	Sampling	Mechanics: given a plot of the samples of $x(t) = \sin(2\pi(3)t)$, determine the sampling period T that was used to obtain them
8	Sampling	Nyquist: given plots of 4 sinusoids, determine which one could be sampled at a rate of 5 Hz without aliasing
9	Trans/Filt	Filtering of a sinusoid: given an LTI system defined by a plot of its freq. response (magnitude and phase), determine the output when the input is a sinusoid
10	Trans	Time/frequency: given plots of a windowed sinusoid and its transform, recognize the transform of a higher-frequency windowed sinusoid given its time plot
11	Conv	Convolution: given plots of the input and the impulse response, recognize the output of an LTI system
12	Trans	Transform properties: given a plot of $P(e^{j\omega})$, find the plot of $R(e^{j\omega})$ when $r[n] = p[n] * p[n]$
13	Trans	Transform properties: given a plot of $P(e^{j\omega})$, find the plot of $R(e^{j\omega})$ when $r[n] = 2p[n]$
14	Conv	Commutative property of convolution: recognize that reversing the roles of the input and the impulse response does not change the output of LTI system
15	Trans	Fourier series: given a plot of a periodic signal, select the equation that best represents it
16	Math	Difference equations: recognize the form of the solution to an LCCDE with sinusoidal forcing
17	Trans	Transform properties and LTI systems: given a system with freq. response $H(e^{j\omega}) = e^{-j\omega\alpha}$ and a plot of the system output $y[n]$, find the plot of the corresponding input $x[n]$
18	Trans	Poles/zeros: given a set of PZ plots, find those corresponding to stable, causal systems
19	Trans	Poles/zeros: given a set of PZ plots, find those corresponding to real impulse responses
20	Trans	Poles/zeros: given a set of PZ plots, find the one corresponding to decaying exponential impulse response
21	Trans	Poles/zeros: given a set of PZ plots, find the one corresponding to a particular frequency response magnitude plot
22	Trans	Fourier transform: given a plot of $x[n]$, determine the DC value of its Fourier transform
23	LTI/Conv	Convolution/LTI properties: given the impulse responses of two systems, determine the causality of their cascade and parallel connections
24	LTI	Linearity/Time Invariance: given 3 input/output pairs, infer whether a system could be linear and/or time-invariant
25	Trans/Filt	Filtering of windowed sinusoids: given plots of a windowed sinusoid, its Fourier transform magnitude, and the frequency response magnitude of a filter, select the plot corresponding to the output

Fig. 1. Table of DT-SSCI (from [11])

Question 1

Figure 2(a) shows four signals $x_a(t)$ through $x_d(t)$, all on the same time and amplitude scale. Which signal has the highest frequency?

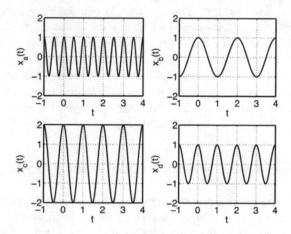

(a) Signals $x_a(t)$ through $x_d(t)$ for Question 1.

Fig. 2. First Question of CT-SSCI (from [11])

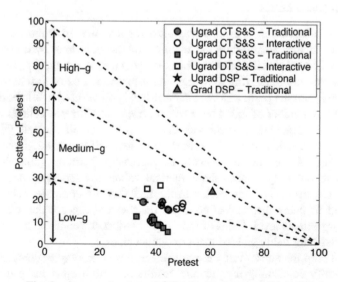

Fig. 3. Averaged gains of different course types (from [1])

3 Robotics Concept Inventory

The robotics concept inventory is divided into 10 categories as summarized in Fig. 4. The categories represent the wide field of aspects that make up the field of robotics. They have been derived from textbooks, curricula and course syllabi as presented in [13]. An initial set of 8 concepts (marked with *) already was discussed there, such that the following table focuses on the newly presented concepts and questions.

The concepts to make up the robotics Concept Inventory and the questions to be used for the CI test have been developed in a participatory process with academia, research, with lecturers and students, as we will show in the subsequent chapter. However, finalization requires a wider verification and possibly refinement. This is the next phase of the Robotics CI development [14].

4 Robotics Concept Inventory Development Process

The Robotics CI development process up to the current point consisted of three phases. In the first alpha-test phase a subset of the questions has been used in selected robotics courses. In a second phase the categories to make up a general robotics concept inventory have been discussed with members of the robotics community. Eventually questions where refined in a game theoretical setup as phase 3. In this chapter we summarize the three phases and show by examples how the phases contributed to the development of the concept inventory and questions.

4.1 Alpha-Test Phase

As a first step, a subset of questions was used for a shortened Robotics CI test. The early feedback helped to improve the wording of the questions by correcting ambiguities and avoiding specific phrases that limited understanding, e.g. by non-native English speakers and improving drawings and representations of formula.

In a second step the partial robotics CI test has been verified and calibrated in pre- and post-tests. A number of initial tests were carried out with two cohorts of the same computer science graduate-level robotics courses. For this, all students jointly participated in the pre-test, 50% participated in the post-test half way through the course and 50% at the end of the course, showing an increasing learning gain. However, the absolute test results only reached the numerical values for undergraduate students, as set by other CI tests. This was related to the fact that the computer science students were exposed to many mechanical concepts for the first time, which would position them at undergraduate level. Gains of the early post-test stayed at the level of low-gain courses, whilst later gains yielded medium-gain values.

In a third step the subset test were applied in a regular way to comparable course setups, generally confirming the previous results and confirming the calibration with existing concept inventories. However, due to ongoing optimisation of the test, the results could not be related on a one-to-one basis. The results of the alpha-test phase of the robotics concept inventory were reported in [13, 15].

#	Category	Concept(s)
1	Math and numerical methods	Reference systems: Identify functional relation between two coordinate systems.*
2		Approximation: Identify approximation for a function that is otherwise hard to compute.*
3		Signal processing: Identify correct time-dependency of a signal.*
4	Mechanics	Basic mechanical principles: Identify accelerated movement from sensor readings.*
5		Basic mechanical building blocks: Identify the correct set of building blocks in mechanics. The question is related to the 3 fundamental mechanical notions of spring, damper and mass with functional relations between forces and motion.
6	Kinematics	Basic kinematic building blocks: Identify the set of basic kinematic components (for stationary robots). The question is related to the 3 fundamental kinematic elements for robots, links and rotational and prismatic joints. Other joints may be built from combinations of the basic joints.
7		Forward kinematic: Identify the work space of a robotic structure. The question is related to combining the parameter ranges of all joints in a robot and deriving the geometric shape, the robot can reach.
8		Inverse kinematic, locomotion of mobile robot: Identify robot configuration to perform a specific movement.*
9	Stability	Center of mass and support polygon: Identify mechanically unstable configuration of a kinematic structure.*
10	Dynamics	Rigid body dynamics: Identify effect of (linear) forces on a rigid body.*
11		Rotational movement: Identify effect of rotational forces vs. momentum. The question is related to the relationship between rotational and linear force and momentum.
12	Control theory	Feedback: Identify the meaning of feedback loops in control systems. The question is related to the calculating the deviation from a desired state (error) by subtracting actual value and set value.
13		Limits (quality / accuracy) of control systems: Identify the control deviations. The question is related speed /accuracy trade-offs in control as apparent by overshots.
14	Sensing	Time in sensing /sampling theorem: Identify the minimum sensing rate to detect cyclic events. This question is related to the sampling theorem.

Fig. 4. Table of robotics concept inventory.

15		Quantification and measurement accuracy: Identify errors in sensor data. The question asks for what kind of actions can be based on sensor data with uncertainty.
16	Perception	Object properties: Identify basic properties of objects. The question is related to properties of objects to be used for landmarks / points of reference.
17		Complex environments: Identify suitable graphical shapes for an occupancy grid. The question is related to appropriately cover an environment, e.g. for mapping.
18	Behavior	Reactive behavior: Identify a suitable configuration for a (simple) reactive mobile robot. The question asks for direct sensor- motor coupling of a differential drive robot for a light-searching application.
19		Sequential behavior: Identify finite state machines (FSM) / automata as a means for sequential activities. The question is related to identify an FSM to detect a specific change in two subsequent data values.
20	Decision-making and planning	Metrics for decision-making: Identify cost / goal functions for robots. The question is related to a metric to base decisions on competing options on.
21		Strategies: identify a suitable strategy to traverse a maze*
22		Complex plans: Identify a suitable path planning approach. The question is related to graph theory.
23		Conflicts: Identify a correct method to resolve conflicts. This question is related to the subsumption architecture in robotics.
24	Uncertainty	Probabilities: Identify effects of uncertainty on localization. The question points towards the direction of Markov localization.
25		Reality/model relation: Identify the reliability of a model in relation to reality. This question points towards Bayes probabilities.

Fig. 4. (*continued*)

4.2   Collecting Community Input

Additional questions have been developed through intense discussions within the scientific community. The robotics education community specifically emphasized the sensing and perception concepts, whilst the artificial intelligence put a focus on decision-making and planning. Experts from the field of autonomous robots considered the concept category on uncertainty as highly relevant.

4.3   Involvement of Students

Eventually graduate-level engineering and computer science students were involved in further development and early testing of individual questions in a game theoretic setup. For this robotics CI questions under test were carefully revised and introduced to exams counting for grading. This way a game situation with pay-off was created. The questions followed the CI-test scheme of one-out-of-four multiple-choice questions. However, some questions slightly emphasized concrete knowledge, related to the lecture over more general concepts as required for CI tests. In addition to trying to answer the question, students had the option to receive partial points for clearly stating that they don't have sufficient knowledge to answer the question. Providing the wrong answer yielded 0 points, 'no answer' 1 point and a correct answer 3 points. Some questions were weighted with a factor of 2 or 3 to reflect the average time to answer. Thus the approach clearly differs from CI questions. However, the approach helped to indirectly involve students in the development of CI questions.

Following are three examples:

Question A: Which geometric shape is not suitable for an occupancy grid?

- Hexagon
- Pentagon
- Square
- Triangle
- No answer

This question actually was introduced as a robotics CI question after observing master-level students performance on this question. On average roughly 5% of master-level students with some robotics background failed, 85% succeeded and 10% decided not to answer the question for lack of knowledge or confidence in knowing the correct answer.

Question B: The following drawing (drawing with links, joints, coordinate systems and measures provided) shows the kinematic structure of the left arm of the Baxter robot (7 degrees of freedom). Given the arm's configuration vector [0 0 0 0 0 0 0] (the angles of the rotational joints), which is the correct transformation matrix between reference frame 0 and 7?

- Matrix a
- Matrix b
- Matrix c
- Matrix d
- No answer

On question B on average roughly 40% of students failed, 5% succeeded and 55% decided not to answer. A similar question, however, asking to apply a concrete 'recipe' to derive the Denavit-Hartenberg table for the kinematic structure of the Baxter robot arm, yielded roughly 20% failures, 25% of correct answers, and the same percentage of 'no answers'. One is tempted to consider the two questions as complicated, but apparently more simple questions like the following question C yielded similar results.

Question C: The Pepper locomotion system uses drives with 6000 RPM, a transmission system 25:1 and a wheel diameter of 14 cm. How high is the approximate theoretical maximum speed?

- 0.2 km/h
- 1 km/h
- 2 km/h
- 6 km/h
- No answer

Interestingly the highest average percentage of about 50% to 60% of 'no answer' was found on a question that required calculations, which shows a low willingness to actually 'work' on answers rather than applying general intuition or knowledge. This observation needs to be carefully watched for the further development of concept inventories and shall be subject to ongoing research.

5 Summary and Future Work

5.1 Summary

In this paper we presented a proposal for a robotics concept inventory and sketched the questions for a test. The full test will be made available to lecturers upon request and evidence of teaching. We also presented the details of the development process and data derived during it. Concept inventories showed to be a powerful yet lightweight tool for assessing and developing courses. The presented robotics concept inventory for the first time allows applying this tool to the field of robotics.

5.2 Future Work

We think, the first results applying concept inventories to the field of robotics are very promising. However, there are some limitations of the current work. To name one prominent limitation: Robotics have become a broad topic – from industrial robotics to health care robotics.

We feel, that our proposed concept inventory is able to cover the fundamentals of robotics, but is not suitable for industrial robotics and health care robotics to the same extent. To give an example: The concept "human robot interaction" is of a higher importance for health care robotics than for industrial robotics. We think that it is necessary to think about (additional) concept inventories for the different areas of robotics.

References

1. Bielaczyc, K., Collins, A.: Learning communities in classrooms: a reconceptualization of educational practice. In: Reigeluth, C.M. (ed.) Instructional Design Theories and Models. A New Paradigm of Instructional Theory, vol. 2. Erlbaum, Mahwah (1999)

2. Winkler, K., Mandl, H.: Learning communities. In: Pawlowsky, P., Reinhardt, R. (eds.) Wissensmanagement in der Praxis. Luchterhand, Neuwied (2002)
3. Aebli, H.: Zwölf Grundformen des Lehrens: Eine Allgemeine Didaktik auf psychologischer Grundlage. Medien und Inhalte didaktischer Kommunikation, der Lernzyklus. Klett-Cotta, Stuttgart (2006)
4. European Commission: The European Qualifications Framework for Lifelong Learning (EQF), Luxembourg (2008), available at http://www.ecompetences.eu/site/objects/download/4550_EQFbroch2008en.pdf. Accessed 05 Feb 2018.
5. Gerndt, R., Lüssem, J.: RoboCup – Wettbewerb als ein Ausbildungskonzept?, Norddeutsche Konferenz für Informatik an Fachhochschulen, Wolfenbüttel (2010)
6. Lüssem, J., et al.: Combining learning paradigms to ensure successful learning outcomes in the area of software development. In: EDULEARN 2011, Barcelona (2011)
7. http://www.foundationcoalition.org/home/keycomponents/concept/index.html. Accessed 05 Feb 2018
8. Crosier, D. Purser, L., Smidt, H.: Trends V – Universities shaping the European higher education area. European University Association, Brussels (2007)
9. European Association for Quality Assurance in Higher Education: "Standards and Guidelines for Quality Assurance in the European Higher Education Area", Helsinki (2009). http://www.enqa.eu/pubs.lasso
10. Vroeijenstijn, T.: A Journey to uplift Quality Assurance in the ASEAN Universities, Report of the AUNP (2006)
11. Wage, K.E., Buck, J.R., Wright, C.H.G., Welch, T.B.: The signal and systems concept inventory. IEEE Trans. Educ. **48**(3), 448–461 (2005)
12. Hestenes, D., Wells, M., Swackhamer, G.: Force concept inventory. Phys. Teach. **30**, 141–158 (1992)
13. Gerndt, R., Lüssem, J.: Towards a robotics concept inventory. In: 6th International Conference on Robotics in Education. Yverdon-les-Bains, Switzerland (2015)
14. Lindell, R.S., Peak, E., Foster, T.M.: Are they all created equal? A comparison of different concept inventory development methodologies. In: PERC Proceedings, vol. 883, pp. 14–17 (2006)
15. Gerndt, R., Lüssem, J.: Concept inventories for quality assurance of study programs in robotics. In: 7th International Conference on Robotics in Education, Vienna, Austria (2016)

Workshops, Curricula and Related Aspects

RobotCraft: The First International Collective Internship for Advanced Robotics Training

Micael S. Couceiro[1,2](✉), André G. Araújo[1], Karen Tatarian[3],
and Nuno M. F. Ferreira[4]

[1] Ingeniarius, Lda., Rua Coronel Veiga Simo, Edifcio CTCV,
3025-307 Coimbra, Portugal
{miguel,andre}@ingeniarius.pt

[2] Institute of Systems and Robotics, University of Coimbra,
Rua Silvio Lima, Polo II, 3030-290 Coimbra, Portugal
micaelcouceiro@isr.uc.pt

[3] American University of Beirut, Beirut 1107 2020, Lebanon
tatariankaren@gmail.com

[4] RoboCorp, Department of Electrical Engineering (DEE),
Engineering Institute of Coimbra (ISEC), Polytechnic of Coimbra (IPC),
3030-199 Coimbra, Portugal
nunomig@isec.pt

Abstract. This paper describes a two-month summer collective internship conceived to provide a unique hands-on experience in robotics. The objective of the Robotics Craftsmanship International Academy, or RobotCraft for short, is to introduce higher education students in the full design cycle of a mobile robotic platform, providing training in computer-aided design (CAD), mechatronics, low-level programming of embedded systems, high-level development using the Robot Operating System (ROS), and artificial intelligence. This non-academic teaching, which successfully completed its second edition, already encompassed around 150 students and 100 universities, being evaluated by participants as challenging, engaging, and beneficial not only to their overall understanding of robotics, but also guiding them through their future academic and professional endeavors.

Keywords: Educational robotics · Collective internship · CAD
Mechatronics · ROS framework · Arduino programming
Artificial intelligence

1 Introduction

Nowadays, robotics is addressed, to some extent, in most engineering courses, including Aerospace, Mechanical, Industrial, Electrical and Computer Engineering, as well as in Computer Science[1]. While a Computer Science programme

---

[1] http://www.thetechedvocate.org/robotics-the-next-big-thing-in-higher-education/.

© Springer Nature Switzerland AG 2019
W. Lepuschitz et al. (Eds.): RiE 2018, AISC 829, pp. 31–43, 2019.
https://doi.org/10.1007/978-3-319-97085-1_3

may focus on the design of artificial intelligence algorithms for image recognition and navigation, a mechanical engineering programme may primarily focus on manipulation kinematics and grasping ability of robotic arms. For students willing to engage in robotics, however, it can be quite challenging to find an introductory all-encompassing training that empowers them with the knowledge to fully develop and experiment with their own autonomous robots [1,2].

This paper describes a training programme, entitled as Robotics Craftsmanship International Academy (RobotCraft)[2], organized as a collective internship. Next subsection presents a brief description of the programme. It is then further delineated in Sect. 2, wherein the objectives and tasks of each module are highlighted. Section 3 summarizes the scores and feedback from the participating students in the programme. Finally, Sect. 4 ends with conclusions and future directions of the programme.

1.1   The RobotCraft Programme

RobotCraft was organized by Ingeniarius, a private company that develops intelligent systems and robotic solutions, in collaboration with the Institute of Systems and Robotics from University of Coimbra (ISR-UC), as an international collective internship with a summer programme in robotics for higher education students. The participants attending this two-month programme have the opportunity to work in robotics within a non-academic setting, focusing on several state-of-the-art approaches and technologies. The programme, which concluded its second edition in the summer of 2017, provides a general overview of the science and art behind robotics, teaching the basis around computer-aided design (CAD), mechatronics, Arduino low-level programming, Robot Operating System (ROS) high-level design, and artificial intelligence. The programme is therefore divided into multiple modules, designated as crafts, carefully prepared to provide a wide range of skills and knowledge in the topic (see Sect. 2 for a more detailed description of each module).

The 2016 version, with a registration cost of 250, brought together 65 students from 35 universities spread over 18 different countries. The 2017 version, with a registration cost of 275, encompassed 84 students from 60 universities coming from 22 countries. Participating countries include Algeria, Egypt, Estonia, Finland, Germany, Greece, Hungary, Italy, Jordan, Kazakhstan, Kosovo, Lebanon, Malaysia, Morocco, Netherlands, Palestine, Portugal, Romania, Russia, Spain, Sweden, Turkey, United Kingdom and Syria. Furthermore, all participants are endorsed in technologically-related fields, namely Computer Sciences, Electrical Engineering, Mechanical Engineering, Aerospace Engineering, Mechatronics, Educational Robotics, Industrial Engineering, and others. As most internships, every participant goes over 8-hour work days from Monday to Friday, divided into theoretical training (1 day a week), mandatory laboratory training (1 day a week), and open laboratory (3 days a week).

---

[2] http://robotcraft.ingeniarius.pt/.

Even though participants were mostly fluent in English, they possessed a highly variable knowledge of other technical skills. Some students barely had any programming experience, while others were already familiar with Arduino, Raspberry Pi, Linux and a very small number (about 2% of the whole number of participants), were familiar with ROS. Therefore, participants were selected based on their speaking, reading and writing English skills, their motivation letter to work in robotics, and their basic knowledge in one of the following topics: programming, CAD, and mechatronics. Nevertheless, to tackle the different backgrounds, a preliminary intensive training in programming and Linux was provided. These additional training hours start one week before the programme and end by the end of its first week, being taught in a laboratory setting with hands-on experience, in both Arduino C programming and introduction to Linux, for an overall duration of 50 h.

Next section dissects the RobotCraft programme by describing the multiple crafts it encompasses.

2   Crafts - Training Modules

As a collective internship conceived to provide a unique hands-on experience in the full design cycle of a mobile robotic platform, RobotCraft is divided into multiple training modules, or crafts, including one purely theoretical craft (Introduction to Robotics), five technical crafts (Computer-aided design, Mechatronics, Arduino low-level programming, ROS high-level programming and Artificial Intelligence), and one experimental craft (Competition). This training is mainly fostered by the private company Ingeniarius, which receives important contributions from ISR-UC and other universities to introduce the theoretical component of every craft. This academic contribution intends to balance the technical skills promoted during the other crafts, with keynote presentations from invited speakers brought every week. Invited speakers share their own vision about robotics, built upon their past experiences. While some of the topics are easily embraced by everyone, such as hardware and software architectures [3], others are controversial, such as emotions in robotics [4,5].

2.1   Introduction to Robotics

This is the most theoretically-driven craft of RobotCraft, which intends to make participants acquainted with the world of robotics, describing its history and evolution by surveying the literature on the topic, as well as presenting mobile robot morphologies, namely sensors and actuators. Additionally, this craft introduces all concepts that are afterwards explored in all other technical crafts, presenting the overall adopted architecture, the hardware that participants will need to integrate, as well as the perception and actuation abilities their robot should be endowed with.

Table 1 presents the learning outcome assessment matrix (LOAM) of the Introduction to Robotics craft, highlighting the overarching goals, desired learning outcomes, teaching methods and assessments adopted [6].

Table 1. Outline of introduction to robotics.

Overarching goal	Desired learning outcome	Teaching methods	Assessments
Introduction to robotics	Understand the state-of-the-art and limitations of the technology, as well as ongoing work and future trends	Lecture/Invited talk	Diagnostic and final test (beginning and end of programme)
Mobile robot morphologies	Ability to identify the functional principle behind different sensors and actuators	Lecture	Oral questioning; Report describing every component of the robot
RobotCraft robot architecture	Ability to understand the functional architecture of a mobile robot, namely to solve the tasks evaluated under the competition	Lecture	Report with a personalized functional architecture

It is in this introductory craft that participants get to know about the upcoming challenges, motivated by the possibility to fully build a mobile robot as illustrated in Fig. 1.

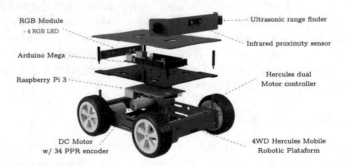

Fig. 1. Illustration of the RobotCraft robot.

2.2   Computer-Aided Design

Still in an early stage of the programme, participants are introduced to CAD with the intent to model the external structure of their own robot. As all other tools adopted in this RobotCraft, the use of open-source or free software for product and industrial design is also promoted, either by resorting to Blender (2016 version) or

Table 2. Outline of computer-aided design.

Overarching goal	Desired learning outcome	Teaching methods	Assessments
Introduction to 3D modeling tools	Get to know the available 3D CAD tools and learn the basis of FreeCAD or Blender	Lecture/Laboratory	Tutorial exercises
Introduction to rapid prototyping	Get to know the available 3D printers and learn how to operate Makerbot Replicator 2	Lecture/Laboratory	Tutorial exercises
Model and print a robot structure	Ability to design and print a robot structure	Laboratory	3D and printed model (measurements)

Fig. 2. 3D models of the robots' external structures designed by participants.

FreeCAD (2017 version). RobotCraft is one of the many programmes contributing towards the acceptability and adoption of open-source and free softwares by the global robotics business community [7,8].

Table 2 presents the LOAM of the Computer-aided Design craft.

This craft provides a higher degree of freedom for participants to perform a more creative task that goes beyond the other engineering-oriented crafts. Every participant is given the 3D model of the major components of the robot (*e.g.*, base platform), as well as the data-sheets of the electronic boards (*e.g.*, Arduino Mega, Raspberry Pi 3, etc.). Considering these constraints, participants can model the external layer of the robot, which will then be printed in polylactic acid (PLA), at will. Figure 2 depicts some of the models developed by the participants under both 2016 and 2017 versions.

2.3   Mechatronics

The Mechatronics craft is perhaps the most hands-on experience of the programme, wherein participants need to fully assemble their robot [9]. Mechatronics deals with every aspect of the mechanical and electronic design of the platform, from screwing the parts together to soldering cables.

Table 3 presents the LOAM of the Mechatronics craft.

Figure 3 depicts one of the hardware architectures adopted during the programme and that participants need to follow and test. Mechatronics laboratories allow participants to get acquainted with a large number of tools, including multimeter, oscilloscope, soldering iron, and others.

Table 3. Outline of mechatronics.

Overarching goal	Desired learning outcome	Teaching methods	Assessments
Mobile robot hardware architecture	Ability to understand and follow a hardware architecture	Lecture/Invited Talk/Laboratory	Tutorial exercises
Mechanical assembly	Ability to assemble all mechanical components of a mobile robot	Laboratory	Robustness analysis
Electrical assembly	Ability to design and print a robot structure	Laboratory	Electric circuit analysis

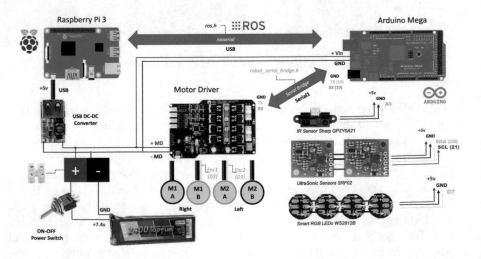

Fig. 3. Hardware architecture of the robot.

2.4   Arduino Low-Level Programming

The Arduino low-level programming is where participants first struggle in RobotCraft. Programming acts as the "glue" connecting all RobotCraft content. In addition, regardless of the attendance of participants in the preliminary training in programming and Linux previously mentioned, most students struggled with coding, understanding the language syntax, libraries, programming paradigms, etc. Therefore, the Arduino low-level programming craft lasts twice the time of the previous crafts, being only exceeded in time by the ROS high-level programming (next section).

Table 4 presents the LOAM of the Arduino low-level programming craft.

Table 4. Outline of Arduino low-level programming.

Overarching goal	Desired learning outcome	Teaching methods	Assessments
Arduino programming	Ability to program Arduino boards to interface with sensors and actuators	Lecture/Invited Talk/Laboratory	Exercises
Kinematics	Ability to develop and implement the kinematic model of a differential drive robot	Lecture/Laboratory	Ground-truth evaluation of real pose
Control	Ability to design and print a robot structure	Lecture/Laboratory	Step-response evaluation of controller

In this craft, participants start by solving multiple exercises using the Tinker-CAD platform[3], then moving to the real hardware, namely the Arduino Mega. Moreover, it is in the Arduino low-level programming craft where many of the concepts presented in the Introduction to Robotics are put into practice. Participants have to explore sensor feedback, mobile robotic kinematics, and motion control [10]. Inside these topics, participants have to carry out several assignments, such as implementing a forward kinematic of a differential mobile robot in order to provide an estimation of its pose over time and a proportional-integral-derivative controller to control both linear and angular speed of the platform over some given reference. Therefore, instead of simply focusing on programming, this craft deals with advanced engineering topics that require a solid theoretical and practical in-class training and as a result demanding more laboratory hours.

---

[3] https://www.tinkercad.com/.

2.5   ROS High-Level Programming

The ROS high-level programming is the most challenging for students and, therefore, the longest craft of RobotCraft. Although participants already come with the basic programming skills acquired from the previous craft, the Arduino low-level programming, this craft deals with a whole new programming paradigm, additionally requiring an in-depth Linux mastery.

Table 5. Outline of ROS high-level programming.

Overarching goal	Desired learning outcome	Teaching methods	Assessments
Introduction to ROS	Ability to work with ROS	Lecture/Invited Talk/Laboratory	Tutorial exercises
Robotic simulation	Ability to implement a new robotic simulation setup	Lecture/Laboratory	Portability of simulation to real world
ROS integration	Ability to integrate the mobile robot in ROS	Laboratory	ROS computation graph and nodes

Although other (easier) frameworks could have been explored, the Robot Operating System (ROS)[4] stood out as the *de facto* standard [11]. ROS is an open-source meta-operating system for robots, providing hardware abstraction, low-level device control, message-passing between processes, and package management. ROS encompasses multiple tools and libraries, constantly updated by the robotic community, to help software developers create robot applications. The number of ROS-enabled robots has been increasing at a fast pace, accounting for hundreds of educational and professional solutions[5].

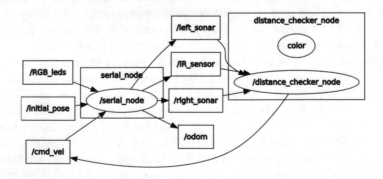

Fig. 4. ROS computational graph provided by `rqt_graph`.

---

[4] http://www.ros.org/.

[5] http://robots.ros.org/all/.

Table 5 presents the LOAM of the ROS high-level programming craft.

Due to the demanding nature of the ROS high-level programming craft, the learning outcomes are assessed more systematically, formally, and quantitatively. Figure 4 depicts an example of the several assignments participants have to deliver throughout this craft.

2.6   Artificial Intelligence

The technical component of the RobotCraft programme ends with the Artificial intelligence craft. This craft further extends the ROS high-level programming craft by focusing on the development of the robot's behavior, which depends on the goal of the competitions in every version of the programme (see next section). The Artificial intelligence craft focuses on simplistic biologically-inspired approaches and finite-state machines (FSM) [12,13]. First, participants have to implement multiple methods, which are evaluated in laboratory context. This is followed by an evaluation using the simulators previously explored and finally using the real robotic platforms.

Table 6 presents the LOAM of the Artificial intelligence craft.

Table 6. Outline of artificial intelligence.

Overarching goal	Desired learning outcome	Teaching methods	Assessments
Introduction to AI	Understand the state-of-the-art and different classes of AI approaches	Lecture/Invited Talk	Oral questioning
Biomimetics & FSM	Ability to theoretically implement AI algorithms for robotics	Lecture/Laboratory	Exercises
Preparation for the RobotCraft competition	Ability to implement and evaluate different AI algorithms for robotics	Laboratory	Algorithmic performance

2.7   Competition

Robotic competitions are known to be well-suited scenarios to experiment, research, and develop new approaches in the several topics surrounding it, including all the crafts addressed in RobotCraft [14]. In the last week of the programme, guidelines are provided to assist participants in the final implementation of the AI algorithms introduced in the previous craft (previous section), as well as fine tuning other features of the robot (*e.g.*, gains of the controllers). The few lectures are then seen as instructional time, primarily spent guiding participants through the processes of the implementation of algorithms and working through the difficulties and pitfalls of real hands-on development.

Despite these technical details, the Competition craft is more of an experimental craft as participants are allowed to test their solutions in test scenarios built inside the laboratories. This was necessary to ensure that every team would be able to complete by the end of the week in a public final competition.

Table 7 presents the LOAM of the Competition craft.

Table 7. Outline of competition.

Overarching goal	Desired learning outcome	Teaching methods	Assessments
Final guidelines and tests	Ability to conclude the full development of a mobile robot	Lecture/Invited Talk/Laboratory	In-lab performance
RobotCraft competition	Ability to test a mobile robot under time-bound real settings	Competition	Out-lab performance

One of the trials under evaluation is the same every year: maze solver. The maze solving is the traditional perpetual competition of RobotCraft, wherein participants need to explore algorithms that allow the robot to navigate until it finds the end of the maze. For this task, participants explore wall-following techniques and other biologically inspired approaches using FSMs. The second competition usually changes every year. While in 2016 participants had to implement a biologically-inspired swarm approach to find the brightest spot within an enclosed environment, the 2017 version brought the possibility for students to implement multi-robot patrolling algorithms using FSMs. Figure 5 depicts an example of the several assignments participants have to deliver throughout this craft.

3   Results and Feedback

To have a clear understanding of the impact of the courses of the second version of Robotcraft and their effectiveness, two surveys were done with the first being at the beginning of the internship and the second during the last week. The purpose of the first questionnaire was to assess the level of skills and knowledge of participants. The results of the latter is based on the reply of 89% of the students, where 25% and 75% were female and male, respectively. The second version of RobotCraft attracted interns of variable academic levels, where the majority, forming 80% of the participants, were students completing their bachelor degrees,

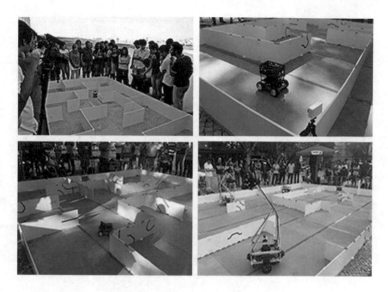

Fig. 5. RobotCraft competitions.

12% were master students, and the remaining 8% had received their masters degree and/or were PhD students.

The results showed that the purpose of enrolling for the majority was to gain experience through an internship followed by the motive to participate in an international event. The primary survey showed that about 40% of the students had a very weak understanding of ROS, followed by AI, and kinematics. On the other hand, mechatronics, control, and 3D printing were seen as less challenging by the majority, with only 20% of participants ranking them as difficult subjects. Finally, CAD was classified as a clearly understood topic by most.

The purpose of the second survey, which was done during the last week of the program, was to evaluate the learning outcomes of courses from the perspective of interns themselves. The results of the latter are based on 72% of the participants. The six main themes and topics of the programme were examined separately. The participants were asked to answer using a 1 to 3 scale, with 1 representing a low level and 3 a high level. Even though many aspects were analyzed in the survey, the herein presented results focus on the understanding of each craft before and after the course (Fig. 6).

One can generally observe an increase of participants' understanding under a given subject during RobotCraft. The ROS subject is worth a mention since it was the most anticipated course for students. As one may observe, the percentage of students ranking their understanding of it as 1, 2, and 3 shifted extremely. While the percentage of participants evaluating their level as low dropped by 42%, the percentage of participants evaluating their level as medium and high increased by 94% and 66%, respectively. This indicates the significant learning outcomes of the ROS course and its matching with the desired goals set by

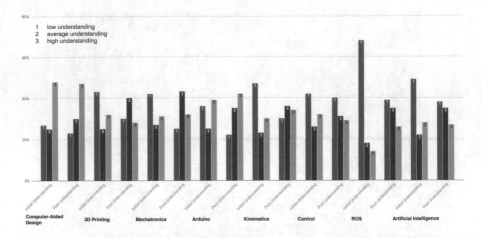

Fig. 6. The initial versus post ranking of understanding of the crafts.

the students initially. Regarding the last craft covering artificial intelligence, it is again noticed the transformations in the level of the skills acquired with the percentage of student perceiving their comprehension of the topic as low, medium, and high decreasing by about 22%, increasing by 60%, and decreasing by 4%, respectively. This later result shows that AI was the only craft in which the number of students who thought to have had a high level of understanding decreased, thus suggesting that many were unaware about the applicability of AI in robotics, finding simplistic methods, such as FSM, to be more challenging than expected.

4   Conclusions

This paper describes the two-month programme denoted as Robotics Craftsmanship International Academy (RobotCraft). RobotCraft was created to encompass students from varying engineering backgrounds and various skill levels, towards the common goal of fully building mobile robotic platforms, including CAD, mechatronics, programming, and AI. The training modules were described, highlighting their overarching goals, desired learning outcomes, teaching methods and assessments adopted.

Based on informal and formal feedback, the training was deemed successful in its ability to provide students an appropriate introduction to a complete robotics design experience. Despite the positive outcomes, several improvements for future offerings are currently being planned, namely to provide a thorough initial training in both C-programming and Linux. Not only those are key requirements of the programme, but also the two competencies wherein students depicted a higher diversity of initial understanding.

Acknowledgement. We sincerely thank all the community for their contributions on the wide range of free and open-source frameworks adopted in this work, namely ros.org, arduino.cc, freecadweb.org, blender.org and ubuntu.com.

References

1. Brady, M. (ed.): Robotics Science, vol. 1. MIT press, Cambridge (1989)
2. Craig, J.J.: Introduction to Robotics: Mechanics and Control, vol. 3, pp. 48–70. Pearson Prentice Hall, Upper Saddle River (2005)
3. Lima, P., Ventura, R., Aparício, P., Custódio, L.: A functional architecture for a team of fully autonomous cooperative robots. In: Robot Soccer World Cup, pp. 378–389. Springer, Heidelberg (1999)
4. Faria, D.R., Vieira, M., Faria, F.C.: Towards the development of affective facial expression recognition for human-robot interaction. In: Proceedings of the 10th International, Conference on PErvasive Technologies Related to Assistive Environments, pp. 300–304. ACM (2017)
5. Cañamero, L.: Emotion understanding from the perspective of autonomous robots research. Neural Netw. **18**(4), 445–455 (2005)
6. Szurmak, J., Petersen, A.: Learning outcomes assessment matrix (LOAM): a software-supported process for identifying and scaffolding complex learning outcomes. Ubiquit. Learn.: Int. J. **2**(3) (2010)
7. Feller, J., Fitzgerald, B., Scacchi, W., Sillitti, A.: Open Source Development, Adoption and Innovation, vol. 10. Springer, New York (2007). ISBN 978-0-387-72485-0
8. Page, T.: Should open-source technology be used in design education? Int. J. Open Source Softw. Process. (IJOSSP) **6**(1), 17–30 (2015)
9. Araujo, A., Portugal, D., Couceiro, M.S., Rocha, R.P.: Integrating arduino-based educational mobile robots in ROS. J. Intell. Robot. Syst. **77**(2), 281–298 (2015)
10. Couceiro, M.S., Figueiredo, C.M., Luz, J.M.A., Ferreira, N.M., Rocha, R.P.: A low-cost educational platform for swarm robotics. Int. J. Robots, Educ. Art, **2**(1) (2012)
11. Quigley, M., Conley, K., Gerkey, B., Faust, J., Foote, T., Leibs, J., Wheeler, R., Ng, A.Y.: ROS: an open-source robot operating system. In: ICRA Workshop on Open Source Software, vol. 3, no. 3.2, p. 5 (2009)
12. Arkin, R.C.: Behavior-Based Robotics. MIT Press, Cambridge (1998)
13. Couceiro, M.S., Vargas, P.A., Rocha, R.P., Ferreira, N.M.: Benchmark of swarm robotics distributed techniques in a search task. Robot. Auton. Syst. **62**(2), 200–213 (2014)
14. Almeida, L.B., Azevedo, J., Cardeira, C., Costa, P., Fonseca, P., Lima, P., Santos, V.: Mobile robot competitions: fostering advances in research, development and education in robotics. In: Proceedings of CONTROLO 2000, The 4th Portuguese Conference on Automatic Control, Guimarães (2000)

University Students Were Creating Activities for Leisure Time Robotic Lessons with Constructionist Approach

Karolína Mayerová, Michaela Veselovská[✉], and Iveta Csicsolová

Faculty of Mathematics, Physics and Informatics, Comenius University in Bratislava, Mlynská dolina, 842 48 Bratislava, Slovakia
{mayerova,veselovska}@fmph.uniba.sk, ivetkacs@gmail.com

Abstract. In this article we describe process, content and participants of the university compulsory elective course "The Robotic Kits in Education 2", which we taught in the winter semester of 2017. We explain specific conditions and circumstances, which encourage transformation of the organization of the course. We focus on analysis of the knowledge our students acquired as the result of this modification. In our research we used the qualitative methods of data collection and data analysis. We compared all students' work and products with the theory of Design of thematic unit and we also present students' conclusions and recommendations for creation of leisure time lessons with robotics.

Keywords: Constructionism · University robotic course
Leisure time robotic activities · Primary and lower secondary school

1 Introduction

Constructionist approach in teaching has many indisputable advantages. The fundamental belief that knowledge is the consequence of experience sets a foundation for the construction of a context for productive learning [1]. However the implementation of constructionist ideas [2] into common compulsory teaching, which is usually unambiguously time restricted, is not very simple. Fortunately, constructionism is already one of the fundamental theories behind Educational Robotics. Educational Robotics provides a learning environment where pupils can interact with their environment and work with real-world problems; in this way Educational Robotics can be an exquisit tool for bringing to the pupils constructionist learning experiences [3]. Robotics can be an efficient tool to get teachers interested in STEM and involved in STEM learning and teaching [4]. Teacher education on robotics can deliver positive impact on teacher practice such as student-centered teaching [5]. Regardless of the significance of teacher education that incorporates educational robotics [6], there are only a few studies that include teacher training with educational robotics [4]. Robotics was utilized in a K-12 teacher development workshop concentrated on helping teachers to

© Springer Nature Switzerland AG 2019
W. Lepuschitz et al. (Eds.): RiE 2018, AISC 829, pp. 44–55, 2019.
https://doi.org/10.1007/978-3-319-97085-1_4

comprehend computer science concepts and programming [7]. But, the workshop did not focus on using robots for teaching. Even when the purpose of teacher training is specifically on the use of robots in teaching, adequate information is rarely provided with regard to how the teacher training was carried out or how its outcome was investigated [4]. Osborne et al. [8] emphasized the significance of teacher training but they mention only briefly the fact that teacher workshops were offered. Arlegui et al. [9] wrote about their teacher training but they provided just anecdotal information on what the teachers learned and did. Tocháček and Lapeš [10] presented a program in which educational robotics was used as a tool to teach pre-service teachers about constructivist teaching. They described more details than several other studies, but they provided no results from the program. There is a scarcity of information and systematic evaluation of teacher education on using robotics for teaching [4].

In this article we describe the process, the content, and the problems which appeared during course *The robotic kits in Education 2*. It is a required elective course within the master's degree program Teacher's Training in Combination of Subjects of Informatics and Mathematics, which follows on elective course within bachelor's degree program called The Robotic Kits in Education 1. This course is elective for programs in teacher's training in all combinations of subjects in our faculty (such as Mathematics, Physics, Chemistry, Biology, Geography and Physical Education). The content of the first course was developed by our colleague Kabatova [11, 12] a couple of years ago. We still use some of her findings to design activities in our course, but we constantly modify the content of it based on the number and characteristics of our students. Therefore we adjusted its content this year too. In our course *The robotic kits in Education 2* we had five students during this school year 2017/2018. Two of them were enrolled in the master's degree programs in areas of Mathematics and Computer Science but also had a part of Teacher's Training. So these two students did not attend didactically oriented subjects at all. However, they still conducted the activities related to teaching. Three of them were enrolled in the master's degree program Teacher's Training in combination of two school subjects. One future teacher (we label her as Jane) was teaching Robotic Lessons in Leisure Time center. In this center Jane could work with many different robotic kits and robots, which we lack in our Department of Informatics Education. This was one of the reasons why we decided to create activities in our course, that would be the most useful for our students and would give them experience with situations where beginner teachers can certainly find themselves.

2   University Course the Robotic Kits in Education 2

In our course we followed the fundamental principles of educational robotics [13], which we used for several years. Based on aforementioned situation we decided to adjust the content of the course *The robotic kits in Education 2* so our students could create activities for Jane's Robotic lessons. Jane brought some robotic kits and robots (which we lack in our department) in our course several times. So at first students could interactively explore new robotic kits and robots and examine the difficulty of working with them and later they created activities for pupils for Jane's Robotic lessons. The aim of our course was to teach students how should they prepare leisure time Robotic lessons for pupils at primary and lower secondary school. During the course students also received verbal and written feedback on created activity from Jane.

2.1 Course Content

Our course lasted 13 weeks (90 min per week). During each lesson we used approximately similar teaching methods. We prepared brief tasks with e-learning system, so students worked mostly independently. They were exploring given robotic kits and robots. They were discussing, distributing tasks and creating activities for pupils. The instructor was acting only as an advisor to students. However, the last lesson was different, because the instructor was conducting group discussion with students and also students were summarizing acquired knowledge. During the semester we discussed the following themes and we discussed some of them for more than one lesson.

1. Introduction to the problematic,
2. Familiarization with robotic kits LEGO Mindstorms NXT and Sphero,
3. Familiarization with robots and robotic kits BeeBOt, AlbiRobot and Code-Pillar,
   – *Creation of first simple activities for them.*
4. Familiarization with robotic kit LEGO WeDo 1.0 and LEGO WeDo 2.0,
   – *Creation of first simple activities for it.*
5. Familiarization with robot Ozobot,
   – *Creation of first simple activities for it.*
6. Testing specific activity with robotic kit LEGO Mindstorms NXT and modifying it for Robotic lessons,
7. Characterizing and defining recommendations and experiences from whole process of creation of activities.

Our department owns only the robotic kits LEGO WeDo 1.0, LEGO WeDo 2.0 and LEGO Mindstorms NXT. Jane was bringing rest of the robotic kits and robots to our course on her own initiative.

2.2 The Participants of Our Course

As we mentioned earlier we had five students (age between 22 and 26 years old) in our course and only one of them was male. He was studying the last year of master's degree program in Applied Informatics. Another student was enrolled in the last year of master's degree program in Mathematics of Economy, Finance and Modeling. These two students did not attend any didactically oriented subjects or pedagogical-psychological subjects, but one of them actively participated in teaching mathematically gifted pupils in her leisure time. Other three students were enrolled in the first year of master's degree program in Teacher's Training in combination of two subjects (one had a combination of Informatics and Mathematics and Jane with last one had a combination of Informatics and Geography). Hence in the last three years (during their bachelor study) they attended several didactically oriented subjects and pedagogical-psychological subjects, where they were learning about preparation for teaching and about several different teaching methods. Didactic subjects included several didactic approaches and methods such as Blooms taxonomy, learning styles or design of Informatics lessons. Besides that students of Teacher's Training attended course The Robotic Kits in Education 1, where they worked with robotic kit LEGO WeDo 1. During our course Jane carried out significant part, because of her teaching of Robotic lessons in Leisure time center, where she worked with two groups that contained of varied numbers of pupils. One lesson at the center with one group took 90 min per week almost every Friday and the groups were consecutive. At the beginning of Robotic lessons pupils were rearranging between two groups. Later there were new pupils coming in, who wanted to attend these lessons. Some of them were pupils from other leisure time activities, that came to visit Robotic lessons and stayed there from 10 to 40 min. The pupils were aged between 8 and 14 (mostly boys) and in one group there were from 2 to 10 pupils. So the numbers of pupils in groups varied every week. During Robotic lessons pupils worked in pairs, in small teams consisted of three pupils or sometimes there were one pupil, who worked alone. These are the same numbers of pupils, which we recommend for one team that was working with one robotic kit based on our previous research with LEGO WeDo at the primary and lower secondary school [14].

3   Methodology

The main goal of our research was to examine, what specific knowledge acquisitions students acquired during our course, since we implemented it using mainly with constructionist ideas. We conducted qualitative research and we used qualitative methods of data collection and data analysis [15]. Our students described designed activities in writing through shared Google document (we briefly describe them in Sect. 4), which we analyzed based on theory about the design of thematic unit [16]. Our student Jane was conducting observations during Robotic lessons, which she taught and she described this observations in writing into shared google document (she also took photos of pupils work and used them in this document). Jane was also describing her findings and conclusions from her observations during our course. We continuously recorded audio of Jane's observations from her Robotic lessons. During the last activities, where students were summarizing acquired knowledge, defining findings and formulating recommendations based on their experiences, we were conducting observations, making field notes and we were recording video of their cooperation and collaboration. Hence participants in our course were conducting action research without knowing it. Almost every hour of our course they were creating new iteration of preparation for robotic activity and they gained feedback from Jane, who also influenced creation of other activities. At the end of course students were describing list of factors, which influenced preparation of content and implementation of Robotic lessons for pupils in primary and lower secondary school. Hence they defined acquired knowledge acquisitions. We describe their findings in Sect. 5. Our students created a theory, which correspond in significant parts with theory about design of thematic unit [16].

4   Activities for Robotic Lessons Created by Students

In this section we briefly describe activities, which were created by our students during course *The robotic kits in Education 2*. After students' familiarization with robotic kits and robots they tried to create activities, which could be consecutive and interesting for pupils. Our students tried to take into account time restrictions and place, where Robotic lessons were implemented. They created set of four activities, which were conducted in five lessons.

4.1   BeeBot, AlbiRobot and CodePillar

The first activity for Robotic lessons contains set of storyline followed tasks with the simplest three types of robots and robotic kits - BeeBot, AlbiRobot and CodePillar (see Fig. 1). Students tried to create interesting narrative taking into account place of Robotic lessons, but they forgot to take into account the time needed to solve the tasks. They also did not prepare solutions of the tasks, although solutions were very important during execution of tasks by pupils. Students did not focus on methodical approach or learning objects of the tasks. Some of the tasks contained lengthy and ambiguous texts. Tasks created with CodePillar had six parts and they were really simple, because it is a educational toy intended for children from three years. Likewise tasks with AlbiRobot had six parts and contained lengthy texts. These tasks focused on familiarization with basic commands. There were several tasks created with BeeBot and they contained even small sketch of two playing areas. However our students did not create real playing areas, so they were missing during execution of the tasks by pupils.

Fig. 1. Pictures from Robotic lesson with BeeBot, AlbiRobot and CodePillar

4.2   Ozobot Evo

Students created two sets of tasks with robot Ozobot Evo in second activity for Robotic lessons. The first set was intended for familiarization with robot and its programming environment. The second set contained narrative, in which pupils should programme robot according to it (see Fig. 2). However there was only one robot so pupils had to take turns. After the lesson the robot had to be recharged. Therefore it could not be used in consecutive Robotic lesson. Nevertheless pupils enjoyed creating a map for the robot (see right bottom of Fig. 2). Even in this activity students did not describe the learning objects or the teaching methods. However Jane additionally created a document with lists of notes about necessary materials, educational methods and so on.

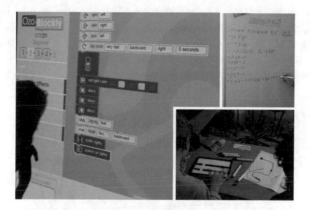

Fig. 2. Pictures from Robotic lesson with Ozobot

4.3   LEGO WeDo 1.0 and LEGO WeDo 2.0

Based on the previous activities and experiences from Robotic Lessons we decided to direct students to think about the didactic approach of the designed activity. So when students were creating tasks for activity with LEGO WeDo, they focused on the organization and structure of teaching, on teaching methods, on the learning content, the learning subjects, the materials and so on. Students did not create particular tasks, but they created a narrative with the use of instructions from the web page education.lego.com (see Fig. 3).

4.4   LEGO Mindstorms NXT

Students were creating this activity during two lessons of our course. So it took them two times longer that creation of other activities. However the robotic kit they were using was more complex than the previous robots, and so the

Fig. 3. Pictures from Robotic lesson with Lego WeDo 1.0 and 2.0

longer time was expected. At first the students were searching for a suitable activity on the web. In the next step they were testing the selected activity and modifying it with the theme about Saint Nicholas. The students designed a sled with reindeers (see Fig. 4). They decided that pupils will build only some parts of the whole robotic model. The other part of this model would be constructed for them. Students even created a solution for the programming task and used a print screen of it in the description of this activity. With the growing difficulty of robotic kits students focused on educational methods, on the learning content and on selecting the appropriate difficulty for the tasks in the activity.

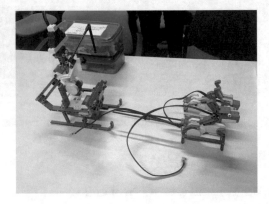

Fig. 4. Robotic model with LEGO Mindstorms NXT created by university students

5   Students Findings

Based on these activities and the observations of our student, who tested these activities with pupils, all our students developed a mind map and several recommendations about what have they learned during whole process. The mind map is divided into two main sections and each section includes four parts (see Fig. 5). The first section (marked with the blue color) relates to the school leadership, that determines each part and where the teacher has little influence. The second section (marked with red color) refers to the overall development of all activities in Robotic Lessons. We analyzed this mind map based on theory of the design of thematic unit [16] and we managed to place all sections from created mind map into several categories from this theory (marked with green color). According to this theory students mentioned almost all its significant parts. However they did not specify pedagogical assessment of the created activities. It could be because in Slovakia leisure time activities are not usually assessed by grading or evaluated at all. Our students mainly focused on preparation of a thematic unit, where they described several general **learning objects** of all activities. They described all **material** needed to conduct every activity including robotic kits with software, computers, office supplies, playing areas, all descriptions of the tasks, the activities for pupils, their solutions for teacher and the rewards for pupils. These rewards were the only feedback from the teacher and its role was mainly to motivate pupils to work. Students also mentioned **place** with regard on rooms' size and equipment. They recommended, that in ideal situation pupils will have opportunity to work in teams on separate tables, with sufficient space for work with robotic kits and a computer. They concluded that **time** has significant influence on Robotic Lessons, because teacher needs to plan all lessons in advance. Teacher should also consider not only the school events and holidays, but also a specific time and day of Robotic Lessons are taught and consider what subjects or activities pupils have attended before and after these lessons. In the part **Organization of work** students focused on how the teacher should prepare before the lessons. The teacher needs to prepare the tasks, have them rigorously tested and have their solution ready. The teacher should also consider how she will motivate pupils and how pupils will be split into groups in a way that they will work effectively and without arguing. Organization of work is influenced by educational **methods**, which teacher can use during the lessons. She could use practical demonstrations and pupils could follow instructions to build or programme robotic model. For example pupils could explore given program and based on this explorations they could work on other tasks. Hence our students recommended to create lessons with regards to the different approaches in teaching. Our students conclude that number of **participants** and their age, gender, nature and skills have significant impact on organization of work during Robotic Lessons. So students speculated about the subject of teaching and about its influence on selection of teaching activities. Our students were even reflecting on the **feedback** from pupils - what pupils thought about the created activities, students thought about differences in performances between the groups of pupils and how they should modify the activities based on their acquired experiences. Hence they naturally incorporated recommendations from the theory about Research of teaching [16], where teacher should reflect on her teaching and try to improve it.

Students also generated several recommendations, which we analyzed and divided into three main categories: Recommendations for organization of work during Robotic lessons, Recommendations for requests of participants and Recommendations for modification of activities.

Recommendations for Organization of Work During the Robotic Lessons. Our students concluded, that the maximum number of pupils in one

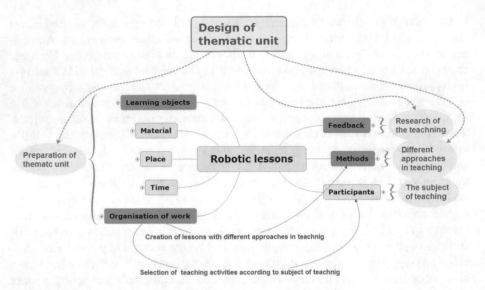

Fig. 5. Mindmap of several factors, which influenced creation of Robotic lessons compared with design of thematic unit

lesson is ten and there need to use several robotic kits of one kind. It is beneficial to combine different types of tasks and to offer pupils a short break between tasks, which take a longer time, otherwise pupils could lose attention, get bored or lose interest in the work.

Recommendations for the Needs of the Participants. During Robotic Lessons pupils often asked about the particular purpose of the created robots or solved tasks from activities in real life. They also needed to know, why they should work on particular activity and what will this work bring to them. Therefore each activity should contain a specific narrative or a tale so pupils will think they are solving something with greater meaning. Older pupils (aged 14) usually wanted to work alone, but they were able to help younger pupils (aged between 8 and 9) a lot. Pupils always asked about a price of robots. Hence pupils realised that robots are not only toys and they should manipulate with them with caution. On the other hand, pupils wanted to have robots at home.

Recommendations for Modification of Activities. Students agreed that activities which they had created should be modified, completed with photos, recommendations and experiences from specific activity of the Robotic Lessons. And if these modified activities appear on blog, they could be used by many other teachers. All activities can be organized by complexity of the types of robots. For example teacher can start with activity with Beebot and finish with activity with LEGO Mindstorms NXT. Activities should be divided based on the types of robots and tasks in a particular activity should specify their difficulty. These tasks should be clear and brief and they should not contain long texts. There should be also tasks where pupils try to create something by using their imagination. For example pupils enjoyed to create something on their own with Ozobot. Finally teacher should always have solutions of all the tasks from activities - mainly programs for robots.

6 Discussion

Students, who are not in the teacher training program, do not usually have any idea how to create tasks for pupils in primary or lower secondary school. They poorly estimate the difficulty of the tasks and the amount of time needed to solve them [17]. We did not come to the same conclusions, although in our course we only had two students, who were not in the teacher training program. But they had experiences with conducting educational activities. However we did observe some differences between them and future teachers. We were sometimes adjusting the process of our course based on appeared circumstances. For example during Jane's' description of observation from her Robotic Lessons, she said she needed the solutions for given tasks, because not only her pupils but even she was not sure, how to solve tasks, which were created by the students who were not future teachers. So we asked them to try to think about the learning objects and didactical approach of the tasks. This situation occurred about half way through the semester, when students began to work with more complex robotic kits and they needed to focus more on teaching methods than on narrative about robotic model.

Besides specific knowledge acquisition related to teacher training, which we verified on created documents and mind map, students acquired other knowledge and skills. These skills relate to how to create modern leisure time activities, which will allow pupils to succeed in 21 century. We assume that our students themselves had the opportunity to develop communication, cooperation and collaboration skills, which are necessary according to [18, 19].

During the last lesson of our course we conducted discussion with all our students. We asked them if they could imagine that they will teach robotics during compulsory school subject Informatics (at elementary or secondary school) or conduct leisure time robotic activities. Only two of them (future teachers) replied, that they will maybe try to teach leisure time robotic activities. Janes' answer was: "Maybe not. It was really hard. There was lack of robotic kits for so many pupils." Female (not future teacher) commented: "It could be fun to try this approach to conduct lessons with robotics. But this theme is so different from everything pupils are learning in school today. It is really hard to create activities, which will be interesting for every pupils in the class. But if I must choose (what to teach) between Mathematics and Robotics, I will choose Mathematics. But if the choice is between Robotics and Chemistry, I will choose Robotics. So it is relative for me." Hence our students were learning how to design and implement robotic activities, they found the whole process very difficult because of many reasons and only few of them could imagine to conduct robotic activities in their future life. So even if teachers recognize the benefits of educational robotics, many of them are not prepared to use robots in teaching [20]. There remains a question, whether work with robotic kits is really so difficult even for students who enrolled in technical programs, or it is based on approach we chose in our course.

In our further research we are going to use several robotic kits in the course The Robotic Kits in Education 2 during the next semester. We recently managed to buy Ozobot, Sphero and Edison. However, there remains another question, whether we should use smaller number of robotic kits in this course, because our students were sometimes frustrated and they found the process of creation of activities very difficult. Frustration could be from a lack of prior experience and confidence in the field, as stated by many pre-service teachers [4].

7   Conclusion

In this article we briefly described university compulsory elective robotic course, where students were learning how to prepare leisure time Robotic Lessons with several robotic kits (BeeBot, AlbiRobot, CodePillar, Sphero, Ozobot, LEGO WeDo, LEGO Mindstorms NXT) and robots for pupils of various age in elementary school. One of our students was teaching these lessons, so that the other students received immediate feedback on the designed activities. We conducted our course mainly with constructionist approach, so our students built their knowledge acquisitions themselves and according to their acquired experiences. We presented a theory created by students at the end of the course, which correspond in significant parts to the theory about design of thematic unit [16]. Students also created several recommendations, which relate to organization of work during Robotic lessons, requests of participants and modification of created activities. Students acquired important knowledge and skills needed to design and implement leisure time robotic activities.

References

1. Stager, G.: Papertian constructionism and the design of productive contexts for learning. In: Proceedings of EuroLogo 2005, Warsaw (2005)
2. Papert, S.: The eight big ideas of the constructionist learning laboratory. Unpublished internal document. Maine, South Portland (1999). Cited in Stager, G.: Papertian constructionism and the design of productive contexts for learning. In: EuroLogo X (2005). http://www.stager.org/articles/eurologo2005.pdf
3. Alimisis, D.: Educational robotics: open questions and new challenges. Themes Sci. Tech. Educ. **6**(1), 63–71 (2013)
4. Kim, Ch.M., et al.: Robotics to promote elementary education pre-service teachers' STEM engagement, learning, and teaching. Comput. Educ. **91**, 14–31 (2015)
5. Bers, M.U.: Blocks to Robots: Learning with Technology in the Early Childhood Classroom. Teacher's College Press, New York (2007)
6. Pittí, K., et al.: Resources and features of robotics learning environments (RLEs) in Spain and Latin America. In: Proceedings of the First International Conference on Technological Ecosystem for Enhancing Multiculturality, pp. 315–322. ACM (2013)

7. Kay, J.S., et al.: Sneaking in through the back door: introducing K-12 teachers to robot programming. In: Proceedings of the 45th ACM Technical Symposium on Computer Science Education. ACM (2014)
8. Osborne, R.B., Antony J.T., Jeffrey F.: Teaching with robots: a service-learning approach to mentor training. In: Proceedings of the 41st ACM Technical Symposium on Computer Science Education. ACM (2010)
9. Arlegui, J., Pina, A., Moro, M.: A PBL approach using virtual and real robots (with BYOB and LEGO NXT) to teaching learning key competences and standard curricula in primary level. In: Proceedings of the First International Conference on Technological Ecosystem for Enhancing Multiculturality. ACM (2013)
10. Tocháček, D., Lapeš, J.: The project of integration the educational robotics into the training programme of future ICT teachers. Procedia-Soc. Behav. Sci. **69**, 595–599 (2012)
11. Kabátová, M., Pekárová, J.: Lessons learnt with LEGO Mindstorms: from beginner to teaching robotics. Group 10, p. 12 (2010)
12. Kabátová, M., Pekárová, J.: Learning how to teach robotics. In: Constructionism 2010 Conference, Paris (2010)
13. Gura, M.: Getting Started with LEGO Robotics: A Guide for K-12 Educators. International Society for Technology in Education, Washington, D.C. (2011)
14. Mayerová, K., Veselovská, M.: Robot kits and key competences in primary school. In: Information and Communication Technology in Education, University of Ostrava, Pedagogical Faculty, Ostrava, pp. 175–183 (2012)
15. Creswell, J.W.: Educational Research: Planning, Conducting, and Evaluating Quantitative. Prentice Hall, Upper Saddle River (2002)
16. Pasch, M.: Teaching as Decision Making: Instructional Practices for the Successful Teacher. Addison Wesley, Reading (1991)
17. Kabátová, M.: Konštrukcionistický prístup vo vyučovaní robotiky v príprave budúcich učiteľov (Constructivist approach in teaching robotics in the preparation of future teachers), FMFI UK in Bratislava, Bratislava (2010)
18. Trilling, B., Fadel, Ch.: 21st Century Skills: Learning for Life in our Times. Wiley, London (2009)
19. Assessment and Teaching of 21st Century Skills: Defining 21st century skills. http://atc21s.org/wp-content/uploads/2011/11/1-Defining-21st-Century-Skills.pdf
20. Mataric, M.J., Koenig, N.P., Feil-Seifer, D.: Materials for enabling hands-on robotics and STEM education. In: AAAI Spring Symposium: Semantic Scientific Knowledge Integration (2007)

Short Course at Brazilian Robotics Olympiad: Forming Competitors

Erika Yanaguibashi[1], Sarah Thomaz[2], and Luiz Marcos G. Gonçalves[1(✉)]

[1] Universidade Federal do Rio Grande do Norte, Natal, RN, Brazil
erikayanaguibashi@gmail.com, lmarcos@dca.ufrn.br
[2] Instituto Federal de Educação, Ciências e Tecnologia do Rio Grande do Norte,
Natal, RN, Brazil
sarah.sa@ifrn.edu.br

Abstract. This paper introduces the successful methodology adopted for the realization of the short course of Educational Robotics offered to the winners of Brazilian Robotics Olympiad's theoretical modality and its results. This short course is free-of-charge hands-on robotics course that is annually offered during the Brazilian Robotics Olympiad national finals of the practical component. The students enter the course without knowledge of robotics and finish the three days course able to join in the practical modality. In the last three years 71 students from high school were awarded with the short course and many of them participated in the practical phase the following year.

Keywords: Educational robotics · Robotics Olympiad · Short course

1 Introduction

The search for new technologies and the great spread of robotics has made this area expand strategically in Brazil on the way to its development. Robotics carries with it a range of possibilities and concepts that can be addressed in classrooms, helping the student in assimilation, understanding, logical reasoning and teamwork. More than that, robotics has been used as a tool for the teaching of transversal contents, such as sciences, physics, mathematics, geography, history and even Portuguese [1].

As Ayorkor Korsahand and Ken Goldberg, founders of the African Robotics Network, said "Robots excite people of all ages. Their physical behavior often inspires primary and secondary student interest in computers, science, math, and engineering" [2].

Due to the great popularization and diffusion of science and technology with society the scientific Olympiads become a great ally for being an incentive for the insertion of new knowledge, previously seen only in courses of higher education or postgraduate. The Brazilian Robotics Olympiad (OBR) [3], on the other hand, has inserted among young people and children the concept of robotics

© Springer Nature Switzerland AG 2019
W. Lepuschitz et al. (Eds.): RiE 2018, AISC 829, pp. 56–64, 2019.
https://doi.org/10.1007/978-3-319-97085-1_5

with the intention of disseminating and engaging them in the area of science and technology [5].

Intended to attract students with or without robotics knowledge, the OBR's activities are divided into two branches, theoretical and practical, the winner of the last one is selected to compete in RoboCup [4] and the winner of the theoretical modality wins a free-of-charge hands-on robotics short course, with duration of 20 h, that is annually offered during the Brazilian Robotics Olympiad national finals of the practical component.

For the students to be contemplated with the course it is necessary that they participate in the theoretical stage, which consists of a test with a set of questions (easy, medium and difficult level) approaching robotics in the contextualization of each item formulated from the contents studied in high school. The tests take place in the student's own city and the best of every state of the country (27 students) are awarded with the course.

The fundamental objective of the short course is to bring students into the realm of robotics, granting the opportunity not only to obtain theoretical content about robots and robotics, but to put into practice all the knowledge acquired in class. As stated by teachers "You will leave here able to participate in the practical competitions of OBR". In this paper we present the methodology used in this short course that enables students who have never had contact with robotics before can, after three days, be able to participate in practical competitions. We also present the experience of the last three years of this short course.

2   Educational Robotics

The increasing search to approach and to introduce new technologies in the classroom gains strength when it comes to the application of robotics in the pedagogical area, aiming to enable an interactive and dynamic learning, with the intention of stimulating a critical thinking resulting in a solution directed to the real world [6]. According to the interactive dictionary of Brazilian education - DIEB [7], the concept of educational robotics is defined as:

> "Term used to characterize learning environments that gather scrap materials or assembly kits composed of various parts, such as motors and sensors controlled by computer and software that allow to program in some way the operation of the assembled models. In educational robotics environments, the subjects construct systems composed of models and programs that control them so that they work in an specific way."

In educational robotics the student begins to construct his own knowledge through his observations and own effort, in this way the learning has much more meaning for him making this a lasting knowledge [8]. Another researcher [9], states that robotics embedded in education came to expand the learning environment. This new feature allows the integration of several disciplines and the simulation of the scientific method, since the student formulates a hypothesis (hypothetical solution), implements, tests, observes and makes the necessary changes so that his robot works [6].

Educational robotics becomes a proposal based on practice (action) and mistakes, inserting a new relationship between teachers and students, in which both walk together in learning, building knowledge and reaching goals [8]. Becoming a playful and challenging activity, which uses a mutual effort between teacher and students, since this one intends to solve a proposed problem, be these composed by hardware and/or software, at the moment in which the teachers use concepts of several disciplines (multidisciplinary) for the construction of this knowledge.

3   Brazilian Robotics Olympiad

In Brazil, the scientific Olympiads began in 1978 and are currently supported by several public institutions. As part of this initiative, in 2007, the Brazilian Robotics Olympiad (OBR), conceived by a team of several university professors, with the main objective of promoting robotics and the contact with science and technology in schools. The philosophy and methodology of the OBR is designed to attract students with or without prior knowledge of robotics, encouraging and opening opportunities for these students to become involved in this new area of knowledge and to know more about associated careers. Among the ten editions that have already occurred in these ten years, OBR records shows that more than 500,000 students have participated in the event. According to 58% of these students, the OBR has influenced the decision of which career to pursue [5].

The activities in the OBR are divided in two modalities, practical and theoretical, each with levels that are designed according to the educational level of each student.

3.1   Practical Modality

The practical modality follows the rules of RoboCupJunior Rescue Line. The mission is characterized by simulating a disaster environment where victim rescue needs to be done by fully autonomous robots. With a team of up to four students it is necessary to complete a task that consists of building a rescue robot that must follow a safe path (black line on a white surface), have to be agile to overcome the hostile terrain (speed reducers) without getting stuck, cross unknown terrain (gaps on the line) where the trail can not be recognized, dodge debris (obstacles) and climb mountains (ramp) to save the victims (balls 2 in. in diameter), transporting it to a safe area (evacuation point) where trained humans can take care of the victim.

By counting with a large number of participants, the practical components of the OBR are divided into three phases: regional, state and national. Figure 1 shows one of the regional phases that occurred in 2016. The robot design, assembly and programming must be carried out by students only, is not allowed teachers to do the tasks for their students, its only possible to them to guide their students on how to solve the problem. There are no restrictions on robots, any

Fig. 1. Regional phase of OBR

kind of material, components and solutions may be used. The winners of the regional phase goes to the state phase and the winners of this go to national finals.

3.2 Theoretical Modality

The theoretical exam of the Olympiad consists of a series of questions that are prepared by a commission made up of teachers from all over the country. The questions are divided into six levels, prepared and distributed according to the students' level of education. Level 0 is for six year old enrolled in elementary school, on the other hand level 5 is for students in the last year of high school. The exams are prepared with the aid of a computer system. Teachers from more than 1,500 Brazilian schools and universities are invited to enter questions through the system.

The exams consist mainly of multiple choice questions and each test is designed to have 25% of the questions easy, 25% complex and 50% medium. The objective of the tests is to disseminate the idea of robotics among students and not to frustrate them, so several questions have a low level of difficulty.

Gold, silver and bronze medals are awarded to the best students of each school, to the best students of each state and to the nation-wide best student. Selected students with the best grades in the theoretical modality that have never had contact with practical robotics are awarded with a free-of-charge hands-on robotics course that is annually offered during the OBR national finals of the practical component.

4 Educational Robotics Short Course

After the result of the best students in the theoretical stage, the 27 approved candidates from each state of the country gain a short course of introduction to robotics. The course is divided into two stages: theoretical and practical. In the theoretical classes, all the necessary knowledge for the understanding of the concepts related to robotics, as well as programming and fundamental concepts for assembly and manipulation of robots are discussed. Theoretical classes are subdivided into the topics presented in the Table 1.

Table 1. Topics of the theoretical classes

Class	Subject
01	What is robotics?
02	What is a robot?
03	Robotics and its applications in the world and market (videos and images to illustrate)
04	Fundamentals of programming logic (variables, expressions and flow controllers)
05	Introduction to R-Educ
06	Introduction to W-Educ
07	Assembling and programming robots

The practical classes are primordial for the construction of this knowledge, in this stage the students are separated into groups of two or three students. Each group receives a robotic kit and a computer. Everything that was exposed in the theoretical classes will be put into practice with the help of teachers, always allowing students to explore and release their knowledge together with creativity to solve each problem proposed in the classroom. The purpose of the course, besides introducing the students in the techno-scientific sphere, aims to develop moral and ethical values such as group work, responsibility and discipline.

On the last day of the course the students are faced with the same problem proposed in the practical stage of OBR: RoboCupJunior Rescue Line. These, without external help, must assemble a robot with the kit received and schedule them to meet the challenge.

5   W-Educ System

With the great diffusion of the educational robotics in the academic environment and the deepening of the researches in this area this research group has been developing, since 2003, several projects allied to the use of the educational robotics environment.

The W-Educ [10] web tool came up to aid in the teaching/learning process of Educational Robotics. This system allows, through the web, that any robotic device to be programmed using the R-Educ programming language. This language is based on a simple and intuitive Portuguese that facilitates the learning of robot programming in a short interval of time.

6   Experiences and Results

After the selection and approval in the theoretical stage, the 27 best students of each state are contemplated with a course of initiation to robotics. The course takes place in the same period of the national phase of OBR (three days) and at all times students are being evaluated with theoretical and practical activities, as well as their participation in classes and teamwork. All these items contribute to the general grade of the student, The student who scores the highest grade at the end of the course acquire an educational robotics kit.

At the beginning of the course the students will know the whole schedule of the classes and how the evaluations will take place throughout the course. On the first day of class, in the morning, students are exposed to concepts about robotics, their laws and applications in the world and market. In the afternoon we begin with the definitions about the fundamentals of programming logic, concept and way of using some electronic components, and then the presentation of the language and the programming environment that will be used.

After presenting all of these concepts, we move on to the next step that consists of the first step of the practical class of the course, for this part of the course we need an educational robotics kit. In the last two years the educational robotics kits used were the Bot'n Roll ONE A, presented in Fig. 2, compatible with the Arduino micro-controller and composed of assembly pieces and electronic components.

Fig. 2. Bot'n Roll ONE A assembled in the course

Step by step the robot is built with the help of the teachers to see if all the components are in the correct places. Because it is electronic components that

Fig. 3. Welding of electronic components

must be welded on a printed circuit board the attention needs to be redoubled so that the entire electrical circuit is not lost due to an incorrect connection.

All the assembly of the mechanical parts, welding of the electronic components and fixation of components was carried out by the students. Of the 11 kits available to students in 2015 only one had its components damaged during the assembly process. We emphasize that one of the prerequisites to be contemplated with the course is to never have had contact with the robotics before. In the Fig. 3 we see a group composed only of women performing the welding of electronic components for the first time in their life. And in Fig. 4 we can see all groups from the 2015 class.

Fig. 4. 2015 robotic class

At the end of the assembly process of the robots, we started the first programming activities with the students, so they began to familiarize themselves with the programming environment used (W-Educ) and the new programming language (R-Educ). The next days of the course are specific to the practice of programming, filled with activities and challenges. With each provided activity we notice that the students are working with more agility and logical reasoning.

On the last day of the course students are prepared to participate in an internal practical competition (among the students themselves), much like the competitions of the practical stage of the OBR, this competition occurs in an official arena, shown in the Fig. 5, however, with a level of difficulty a little lower than the OBR, only because of the limitations of the robot used. This step also counts as a final assessment for student grades.

Fig. 5. Competition arena

At the end of the mini-course each student answers an evaluation questionnaire about the W-Educ system used during the course (the questionnaire does not count towards the student's final score), as a way to evaluate the system by providing feedback to the developers of the environment, as well as an analysis of students achievements. One of the analyzed questions was: "Do you learn with the system?". Among the students who answered the questionnaire, 97.8% answered yes. This data proves that the students learned from the system used, and more than that they learned to program their robot. About this, one of the students said:

"This system is incredible, in just three days I was able to build and program my robot."

In addition, of the 71 students who participated in the last three editions of the course, 68 students successfully completed the final stage of the course with the final internal competition, representing a total of 94.4% success.

Since most of the students who participate in the short course are in the last year of high school they can not participate in next year's practical phase, being necessary, if they wish, join others robotics competitions. However, some of the students who were not in the last year of high school had the opportunity to compete in the practical stage of the following year OBR. One of these students, in the year of 2015 was contemplated with a kit of robotics and founded the first robotics group of his city. This student, in 2016, was state champion and participated in the national finals.

7   Conclusion

Brazil has contemplated the advance of scientific Olympiads, especially robotics, which has shown exponential growth. Every year, this Olympiad, which has theoretical and practical stages, gives the best student in each state a free-of-charge hands-on robotics course.

As presented in this article, this course has enabled students who never had contact with robotics to become future competitors of OBR practical phase and others robotics competitions. According to the data presented, the methodology proposed for this short course can easily be replicated in other proposals, allowing more students to take advantage of the benefits of robotics in the education process.

References

1. Almeida, M.A.: Possibilidades da robótica educacional para a educação matemática. Curitiba (2007)
2. African Robotics Network (AFRON) Internet Site. Developed by DreamOval (2013). http://robotics-africa.org/. Accessed 19 Dec 2016
3. Brazilian Robotics Olympiad Internet Site (2018). http://www.obr.org.br/. Accessed 23 Mar 2018
4. Robocup Federation (2018). http://www.robocup.org/. Accessed 24 Mar 2018
5. Aroca, R.V., Pazelli, T.F., Tonidandel, F., Filho, A.C., Simes, A.S., Colombini, E.L., Burlamaqui, A.M., Goncalves, L.M.: Brazilian Robotics Olympiad: a successful paradigm for science and technology dissemination. Int. J. Adv. Robot. Syst. **13**(5), 1729881416658166 (2016). https://doi.org/10.1177/1729881416658166
6. Nascimento, F.M.S., Santos, F.L., Bezerra, R.M.S.: REDUC: A robótica Educacional como abordagem de baixo custo para o ensino de computação em cursos técnicos e tecnólogos. Instituto Federal de Educação, Ciência e Tecnologia da Bahia (IFBA), Salvador (2013)
7. Dicionário interativo da educação brasileira: Educa Brasil 2012 (2012). http://www.educabrasil.com.br/eb/dic/dicionario.asp. Accessed 10 Dec 2013
8. Maisonnette, R.: A utilização dos recursos informatizados a partir de uma relação inventiva com a máquina: a robótica educativa. Proinfo – Programa Nacional de Informática na Educação – Paraná (2002)
9. Besafe: A casa do Cyberbox (2003). www.cyberbox.com.br.. Accessed 19 Dec 2016
10. Sá, S.T.L., Fernandes, C.C., Yanaguibashi, E.A., Barros, R.P., Burlamaqui, A.M.F., Gonçalves, L.M.G.: W-Educ: a complete, dynamic, web-based educational robotics environment for attracting students to robotics and engineering. Int. J. Eng. Educ. (2017, submitted)

Case Study on Physical Computing with NodeMCU on Summer School

Eva Klimeková[1(✉)], Marek Mansell[2], Karolína Mayerová[1],
and Michaela Veselovská[1]

[1] Faculty of Mathematics, Physics and Informatics,
Comenius University in Bratislava, Mlynská dolina, 842 48 Bratislava, Slovakia
{klimekova,mayerova,veselovska}@fmph.uniba.sk
[2] Faculty of Informatics and Information Technologies,
Slovak University of Technology in Bratislava, Ilkovičova 2,
842 16 Bratislava, Slovakia
marek.mansell@gmail.com

Abstract. This paper introduces our case study on using the NodeMCU device in educational activities with lower secondary school pupils aged between 8 and 13 years. Readily available at a cost acceptable for even small schools, our research is into how such activities can be integrated into the curriculum and describe the hardware package, environment and the four graded activities that we developed. We come to the conclusion that while this approach is not suitable for younger pupils, it did motivate pupils to continue programming and experimenting with physical computing.

Keywords: NodeMCU · Lower secondary school · Physical computing

1 Introduction

Physical computing in primary and secondary schools takes advantage of robots, physical computing devices and constructionist toolkits in education, where researchers tend to focus on a variety of positive aspects of physical computing, such as motivation [1], creativity [2] and cognitive load [3]. Slovak lower secondary education usually utilizes programmable robots like Bee-Bot and Lego Mindstorms [4]. In higher education we encounter more advanced technology, such as Raspberry Pi, Arduino and NodeMCU, which are commonly used in specialised technical courses. The Python community has developed documentation for NodeMCU [5, 6] that is regularly used in introductory physical computing workshops organised by the community. Experience gained from previous workshops has suggested the possibility that this more advanced hardware could, in certain circumstances, be used in lower education, reducing the number of different platforms used by teachers. As organizers of a summer school for pupils aged 8 to 13, we decided to conduct a case study with NodeMCU, where we monitored pupil's reactions to activities with this device. Due to the fact we had no prior knowledge of the pupil's programming skills we prepared activities for computing novices.

Graphical programming environments based on command blocks are key instruments in teaching programming basics to young aged pupils. Among the advantages of

© Springer Nature Switzerland AG 2019
W. Lepuschitz et al. (Eds.): RiE 2018, AISC 829, pp. 65–70, 2019.
https://doi.org/10.1007/978-3-319-97085-1_6

these graphical environments are that they are more intuitive [7, 8], leave learners with a higher feeling of satisfaction [9], therefore strengthen their motivation and self-efficacy [10, 11]. For the described workshop we prepared a Blockly environment and designed several graduated activities.

2 Hardware and Software

The NodeMCU development board is built around the ESP8266 WiFi System on Chip (SoC) [5]. An inexpensive, powerful (32bit) and completely self-contained micro-controller development platform, capable of running MicroPython, and having both WiFi and USB for connectivity, it requires no expensive additional hardware programming devices, making it very suitable for use in a financially constrained educational environment. While it's primary deficiency is it's limited number of GPIO (General Purpose Input/Output) pins, it loses nothing in utility for introductory physical computing courses. Despite this constraint, the advanced architecture of the 32bit processor makes it very suitable for computing courses at higher levels of education.

To support the NodeMCU and provide the physical component of the workshop we selected the following items:

- USB micro cable to program and provide power to the NodeMCU
- Solderless Breadboard to simply create connections
- Jumper Wires
- Red LED with current limiting resistor
- LED strip consisting of eight individually addressable multi-colour LEDs.

Circuits can be constructed using the jumper wires and solderless breadboard to connect the NodeMCU to the other components without the inconvenience and potential burn hazard of soldering, making it safer for younger pupils. Circuits can also be rapidly modified for experimentation and to correct mistakes.

The red LED (with current limiting resistor to prevent it burning out) is used as a basic introduction to configuring microcontroller GPIO pins.

The multi-colour LED strip can be controlled from a single GPIO pin and encourages the pupils to experiment with code since it produces immediate graphical feedback. It does not require current limiting resistors, since they are already built in.

In lieu of an educational software environment (Fig. 1) suitable for our workshop, we developed the ESPBlocks IDE [12], a simple GUI Python application primarily consisting of a HTML browser widget that opens a specific website with a version of the JavaScript Blockly application modified to generate MicroPython code to run on the NodeMCU platform. Once prepared, the pupils can simply click the 'run' button and the environment opens a serial connection with their device and uploads the code via a USB cable.

While there are numerous options for loading code to the NodeMCU, we felt that developing a simplified environment was more suited to the educational level of the pupils attending the workshop. While it was never intended that the IDE be developed further, findings from use suggest it could be further developed as a plugin for use with common development environments and code editing suits.

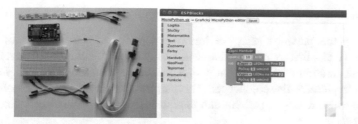

Fig. 1. Hardware kit on the left: board NodeMCU, breadboard, jumper cables, USB micro cable, single colour LED, resistor and an individually addressable LED strip. On the right: Environment ESPBlocks with a code for blinking a LED

3 Methodology

The main aim of our research was to examine how pupils reacted to set activities. A case study was chosen as the main strategy and we used qualitative methods of data collection and data analysis [13], such as observation (transcriptions and field notes) and audiovisual materials (their work by recorded video and the results of their work by photograph).

Our research was conducted during a summer school in Bratislava through the use of a workshop of approximately 90-min duration. Sixteen pupils (7 girls, 9 boys), aged between 8 and 13, were allowed to form into pairs of their own volition. Most pairs were single gender. 3 pupils had no prior programming experience while the remaining 13 had programmed in Imagine Logo or Scratch. Six pupils had some experience with Lego Mindstorms, however, we had no further information about the extent of their knowledge. Every pair of pupils was provided with a hardware kit (Fig. 1, left), a notebook with the ESPBlocks IDE installed (Fig. 1, right), a 'cheat-sheet' for the device and were faced towards the projector and the lecturer.

During the workshop there were a lecturer and three researchers who were collecting data. One of the researchers assisted the lecturer. There was also a teacher helping pupils to montage the hardware.

4 Activities

These are the four activities which we conducted during our workshop.

4.1 Connect the Device and Light the LED

The goal of this task was to familiarise the pupils with the hardware, the ESPBlocks environment and initial concepts of physical computing. Under the instruction and supervision of the lecturer and researchers, the pupils connected the required hardware components and programmed the device to illuminate the LED. This hardware configuration would be used for the next two tasks.

4.2   Blink the LED

Building on the previous task, we introduced pupils to the concepts of loops by blinking the LED, first four times, then a thousand times. We then interactively covered the rate and duration of illumination with a question and answer session before having the pupils implement these concepts. The purpose of this activity was that pupils be able to recognize recurring patterns and to repeatedly apply generalised solutions based on those patters.

4.3   Blink the S.O.S. Sign

When pupils completed the previous task, we set them the task of using this knowledge to make the LED transmit an SOS signal in Morse Code. However, most were unfamiliar with both Morse and SOS which necessitated further explanation. Since different groups arrived at this point at different times, this needed to be done more than once.

4.4   Addressable LED Strip

The final activity taught involved using the addressable LED strip with which, under supervision to prevent damage through misconnection, the pupils replaced the single colour LED and its associated current limiting resistor. Once instructed on changing the colours of individual LEDs on the strip in software, the pupils were allowed free time to experiment with the device (Fig. 2).

Fig. 2. Photo of a pupils work during activity with addressable LED strip and a solution of a task in environment ESPBlocks

5   Findings

Slovak pupils are not introduced to electric circuit theory until they are 14–15 years old (9th year of lower secondary school) when it is introduced as part of the Physics component of the National Educational Curriculum. Subsequently younger pupils did not possess the requisite electronic knowledge for constructing circuits, they did not understand the physical background of the activities. As one pupil said "*I wonder why it has to be joined in the given way, but I do not understand it and it makes me angry*". This strongly suggests that physical computing education should be **synchronized**

with the physics curriculum. In our opinion, activities with hardware configuration should be only introduced when pupils have gained the requisite academic and technical foundations, thus constraining the age at which the subject can be introduced, which in Slovakia would currently be 14. Consequently, we would recommend that pupils younger than 14 should be taught using preconfigured **hardware that does not require knowledge of physics**, such as the Micro:bit [14].

Attention span should also be taken into consideration. The process of the device component connection and programming the LED to illuminate lasted approximately half an hour, which was **too long for younger pupils** who didn't comprehend the underlying physical concepts. As the pupils lacked the knowledge of electric circuits, all of the connections had to be checked before continuing. This concludes that a single teacher, without assistance, cannot manage to check each pair's work before connecting the device to the computer on a 45 min long lesson. In the case of these activities, they would need to reserve more time than a single lesson.

The pupils' work in the environment was also observed and it led us to the conclusion that the blocks should be rearranged and sorted by use in order to reflect the usage in activities. An error was noticed where all of the commands present in the program body were run even if not connected to the main block. The fact that pupils had to work with a touchpad instead of a mouse also caused some difficulties.

The four youngest pupils left before completing the workshop; one had some experience of Lego Mindstorms while the two youngest had never programmed before.

The workshop was rated as *"cool"*, *"great"* and *"interesting"*, because *"I learned new things I liked"*, *"I learned something different"* or *"Because it was fun"* in a questionnaire completed by the pupil who completed the workshop. Despite lacking knowledge on electrical circuits and components, the consensus was that they loved activities with the LED strip the most.

Consequently, based on the pupils' reactions during the workshop, we believe that pupils who completed the workshop were excited, with those less enthusiastic about hardware configuration still excited by their experiences programming the LED strip. In fact, their enthusiasm was such that we skipped the planned final activity with buzzers completely. Subsequent to the workshop, many parents confirmed our belief that their children enjoyed the experience and we believe that many were motivated to continue programming and experimenting with physical computing.

While we wanted to show that the NodeMCU would be suitable for younger pupils, the frustrations caused to some by having to connect components before programming them suggests that an already preconfigured device, such as the Micro:bit, would be more suitable for engaging younger and less technically inclined pupils. However, this means teachers need a second platform for younger pupils, which is the opposite of what this research tried to prove.

6 Conclusion

This paper comprises our case study on using the NodeMCU device in activities for lower secondary school pupils aged between 8 and 13 years. Whereas we are encountering this device more and more recently, and it is affordable even to smaller schools, we decided to conduct a study how such activities can be integrated into the curriculum and describe the hardware package, environment and the four graded activities that we developed. We come to the conclusion that while this approach is not suitable for younger pupils, it did motivate pupils to continue programming and experimenting with physical computing.

References

1. Kaloti-Hallak, F., Armoni, M., Ben-Ari, M.: Students' attitudes and motivation during robotics activities. In: Proceedings of the Workshop in Primary and Secondary Computing Education, pp. 102–110. ACM (2015)
2. Sentance, S., Waite, J., Yeomans, L., MacLeod, E.: Teaching with physical computing devices: the BBC micro:bit initiative. In: Proceedings of WiPSCE 2017, Nijmegen (2017)
3. DesPortes, K., Anupam, A., Pathak, N., DiSalvo, B.: BitBlox: a redesign of the breadboard. In: Proceedings of the 15th International Conference on Interaction Design and Children, pp. 255–261. ACM (2016)
4. Veselovská, M., Mayerová, K.: LEGO WeDo curriculum for lower secondary school. In: International Conference on Robotics and Education RiE 2017, pp. 53–64. Springer, Cham (2017)
5. NodeMCU MicroPython documentacion. https://www.micropython.sk
6. MicroPython. http://naucse.python.cz/lessons/beginners/micropython/
7. Endoh, H., Tanaka, J.: Integrating data/program structure and their visual expressions in the visual programming system. In: Computer Human Interaction, pp. 453–458. IEEE (1998)
8. Neag, I.A., Tyler, D.F., Kurtz, W.S.: Visual programming versus textual programming in automatic testing and diagnosis. In: AUTOTESTCON Proceedings, 2001, IEEE Systems Readiness Technology Conference, pp. 658–671. IEEE (2001)
9. Booth, T., Stumpf, S.: End-user experiences of visual and textual programming environments for Arduino. In: International Symposium on End User Development, pp. 25–39. Springer (2013)
10. Armoni, M., Meerbaum-Salant, O., Ben-Ari, M.: From scratch to "real" programming. ACM Trans. Comput. Educ. TOCE **14**, 25 (2015)
11. Neutens, T., Staes, J., Wyffels, F.: Implementation and evaluation of a simulator and debugger for physical computing environments. In: Proceedings of WiPSCE 2017, Nijmegen (2017)
12. ESPBlocks. https://github.com/marekmansell/ESPBlocks
13. Lichtman, M.: Qualitative Research in Education. SAGE Publications, Thousand Oaks (2013)
14. The Micro:bit. http://microbit.org/

MOOC on the Art of Grasping and Manipulation in Robotics: Design Choices and Lessons Learned

Maria Pozzi[1,2(✉)], Monica Malvezzi[1,2], and Domenico Prattichizzo[1,2]

[1] Department of Information Engineering and Mathematics, University of Siena, Siena, Italy
{pozzi,malvezzi,prattichizzo}@diism.unisi.it
[2] Department of Advanced Robotics, Istituto Italiano di Tecnologia, Genova, Italy

Abstract. This paper presents guidelines for designing a MOOC on Advanced Robotics topics based on the authors' experience in creating an on-line course on *The Art of Grasping and Manipulation in Robotics*. After a revision of the main MOOCs and video lectures about Robotics that are available on-line, we present our course and describe the preliminary feedback we gathered from a group of Master students.

Keywords: MOOC · Robotic grasping · Teaching robotics

1 Introduction

Starting from 2012, MOOCs (Massive Open Online Courses) became fundamental resources in higher education and are now opening new entrepreneurial possibilities [27]. They boost the self learning of students, with easily retrievable and well structured information, and create international educational networks. A relevant advantage of MOOCs is that they make available also very specific courses, that a student couldn't get with traditional learning means. These aspects make MOOCs particularly suitable also for teaching Robotics related topics [5]. The absence of a direct contact between the teacher and the learner in MOOCs may lead to uncertainties on what is actually learnt, so this type of resources needs to be carefully designed and organised [19].

In this paper, we present a MOOC on *The Art of Grasping and Manipulation in Robotics*, that explains the mathematical foundations of grasping and the features of SynGrasp, a MATLAB® Toolbox for the simulation of human and robotic hands.

Providing a robotic system with the ability of reaching, grasping and manipulating objects is one of the main challenges in robotics research. However, to the best of our knowledge, there exist few freely available educational resources regarding this topic, especially for autonomous learners. Robotic grasping and manipulation deserve a special attention also in higher education, so that students and engineers can approach the new challenges in this field with a solid

© Springer Nature Switzerland AG 2019
W. Lepuschitz et al. (Eds.): RiE 2018, AISC 829, pp. 71–78, 2019.
https://doi.org/10.1007/978-3-319-97085-1_7

theoretical background and also the knowledge of the latest results. A MOOC containing the mathematical bases and some of the latest results in the field of robotic grasping could interest students and researchers with various backgrounds, from mechanical and mechatronic engineering, to computer science, neuroscience, and medicine. Indeed, the mathematical models underlying robotic hands can be applied also to human hands and prostheses.

The rest of the paper is organized as follows. In Sect. 2 we analyse available educational resources on robotic grasping and manipulation, and list some of the most important online courses on Advanced Robotics. Section 3 presents our course and describes how our design choices were evaluated by 26 students. The results of the questionnaire were encouraging and provided us with several insights on how advanced robotics topics can be tackled with a MOOC.

2   Previous Work

2.1   Educational Resources on Robotic Grasping and Manipulation

Classical theory and applications of robotic grasping have been summarized by Prattichizzo and Trinkle in the chapter entitled *Grasping* included in the Springer Handbook of Robotics [18], but also in books [13,17], and reviews [2]. In the MOOC that we describe in this paper, we explain the theory of grasping with the help of simulations performed using the SynGrasp MATLAB Toolbox [16], similarly to what was done by Peter Corke in his book [4] and in the MOOCs *Introduction to Robotics* and *Robotic Vision*, where the explained theoretical concepts were associated to MATLAB examples using the Robotics Toolbox [5].

SynGrasp has been developed by the Siena Robotics and Systems Laboratory since 2012, and contains more than 300 functions and scripts for grasp analysis. It has been downloaded more than 4000 times, and it is currently used in 3 European Projects and continuously updated. Thanks to its simplicity and intuitiveness, the toolbox is well suited for education and is already used in the courses of Robotics, Human Centered Robotics, and Mechanical Systems of the University of Siena.

2.2   On-Line Courses on Advanced Robotics

Robotics is starting to face more human-centered problems [24] and is opening many new challenges [1]. This explains why on-line courses on advanced robotics continuously increase in number, and involve some of the best Universities in the world.

One of the most popular robotics courses on YouTube is *Introduction to Robotics* by Professor Oussama Khatib [14]. It was recorded in 2008 during the CS223A course of the Stanford Computer Science Department. The 16 lectures cover all basic topic of robotics and last between 58 and 77 min. The first three lectures had $504, 829$, $176, 534$, and $81, 804$ views, respectively. The average

---

[1] www.therobotreport.com/10-biggest-challenges-in-robotics/.

number of views of the other videos is around 35,000. Another playlist that is on-line since 2008 is the *Lecture Series on Robotics* by Amarnath [1], whereas *Underactuated Robotics* by Russell Tedrake is on YouTube since 2010 [26].

More recent resources include *Robotics 1* by De Luca (2014) [9], *Evolutionary Robotics* by Bongard (2016) [3], *Fundamentals of Neuromechanics* by Cuevas (2016) [7], and *Programming for Robotics (ROS)* by Fankhauser *et al.* (2017) [12]. In 2017, Park and Lynch published their book *Modern Robotics: Mechanics, Planning, and Control* [15], that is enriched with more than 90 videos covering all the chapters of the book[2].

The MOOC entitled *Control of Mobile Robots*, delivered by Egerstedt in Coursera since 2013 [11] was one of the first MOOCs on robotics and was used in a flipped classroom experiment to demonstrate its efficacy [8]. In February 2016, Coursera presented its first Robotics Specialization, consisting of a series of six courses from University of Pennsylvania [6]. Recently, also edX launched several robotics micromasters and courses[3], created by top universities, including Columbia University, University of Pennsylvania, MIT, and ETH Zurich. At the beginning of 2017, Siciliano launched his MOOC on *Robotics Foundations I - Robot Modelling*, that covered some chapters of his textbook [23] and was delivered through the Federica.EU portal.

3 MOOC on *The Art of Grasping and Manipulation in Robotics*

3.1 Design

The act of grasping and manipulating tools is the ultimate interface of a robotic system with the environment and it is one of the most complex tasks in industrial, service and humanoid robotics. This is why it has attracted the interest of the robotics community in recent years, as shown by the increasing number of articles, workshops [20,21,25], and projects on this topic. The development of robotic hands, in particular, is a cutting-edge research field, and the new trend is building soft and underactuated devices that can easily interact with the environment and with humans [10,22]. This new manipulation paradigm is posing stimulating problems from the control, actuation and sensing points of view. Undergraduate students as well as researchers in robotics must have the tools to address these and other challenges that are arising in the field of grasping and manipulation, and for this reason we decided to create a MOOC containing the basic notions about these subjects.

The main challenge for an educator that creates a MOOC is to avoid that students get "lost in information". The learning flow must be clearly stated from the beginning, and video lectures must explain one, or maximum two, important concepts at a time.

---

[2] hades.mech.northwestern.edu/index.php/Modern_Robotics_Videos.

[3] https://www.edx.org/course?search_query=robotics.

The book written by Murray et al. [17] in 1994 is a complete and fundamental reference textbook for the study of robotic manipulation. However, it can be very complex for students without a background in robotics or mechanical engineering. To encourage self-learning and create educational resources suitable for a diverse public, we adopted a pyramidal course structure with three main levels of learning (Fig. 1). We called them "Surfing", "Snorkeling", and "Scuba Diving" to transmit the idea that level by level, students will get a deeper and deeper understanding of the subject. People who have never studied robotic grasping will start by scratching the surface of the topic through very concise lectures explaining basic concepts (Level 1: "Surfing"). These concepts will be then examined in depth by looking at the underlying math, with equations and rigorous proofs in Level 2: "Snorkeling". Level 3: "Scuba Diving" will allow students to apply the knowledge acquired in previous levels to code simulations with the SynGrasp MATLAB Toolbox [16].

The MOOC on *The Art of Grasping and Manipulation in Robotics* was recorded during real lectures and is structured in 4 units: the first belongs to Level 1, the second and the third to Level 2, and the last to Level 3 (see Fig. 1). Unit 1 explains basic notions for understanding robotic grasping, including the difference between power and precision grasps, the friction cone, and the Grasp Matrix. Units 2 and 3 explain the mathematical model of a grasp and how it can be used to design proper control strategies for grasping tasks. Unit 4 introduces the features of SynGrasp Toolbox, and proposes some simulation exercises to the students.

Instead of relying on a specific MOOC platform, we published the MOOC in YouTube Playlists that can be retrieved from the website of the course[4]. This choice has the main drawback that students can't get a certificate after the course, but guarantees a wider spread of it.

Level	Contents
Level 1 ("Surfing")	Unit 1: Basic concepts
Level 2 ("Snorkeling")	Unit 2: Grasp modelling
	Unit 3: Grasp control
Level 3 ("Scuba Diving")	Unit 4: SynGrasp

Fig. 1. Levels of learning (left) and how they are implemented in our MOOC (right).

---

[4] sirslab.dii.unisi.it/MOOC/index.html.

3.2   Students' Feedback and Lessons Learned

Between November and December 2017, the MOOC on *The Art of Grasping and Manipulation in Robotics* was used as a support for the first part of the course on *Human Centered Robotics* held by Prof. Domenico Prattichizzo at the Master of Science in Computer and Automation Engineering at the University of Siena. We asked 26 students, of which 18 declared that *Human Centered Robotics* was their first course in robotics, to fill in a questionnaire about the MOOC. The form was divided into 5 sections and contained a total of 28 questions. Most of them were five-level Likert items, others required a short answer. We report in this section the relevant results, as well as the lessons learned from them.

When asked to rate their satisfaction about the course with a score from 1 ("Not very") to 5 ("Very much"), the majority of students answered with 4 (9 students, *i.e.* 34.6% of the answers) or 5 (8 (30.8%)). Nobody answered 1. Only one student answered 2, and the main concern that he underlined was a lack of exercises in the video lectures. This weakness was underlined also by other 5 students when answering to the question "What is the thing you liked least about the course?", and by the answers to questions C3 and C6 (Fig. 2).

Results summarized in Fig. 2, show that in general students appreciated the MOOC website and contents. Videos were found interesting and were not perceived as too long. One of the questions requiring a short answer was: "What is the thing you liked most about the course?". Students' answers indicate that they especially liked the fact that video lectures can be watched also by student-workers. Most of the students agreed that having video lectures also for other courses would be very useful, and that watching the videos gave them a strong help while studying for the exam.

Interviewed students had different backgrounds, ranging from information engineering and computer science (21) to management engineering (3), physics (1), and mechanical engineering (1). We asked them if their previous knowledge was sufficient for understanding the course: 7 ($\sim$27%) disagreed and expressed the need for additional material on robot kinematics and dynamics and linear algebra.

The results of the questionnaire underlined that enriching university courses with a MOOC can be useful for students, above all when they have to learn specific topics for which educational material is not always easy to consult. We also gathered important suggestions on how to improve our course before promoting it in the robotics community. We will enrich it with more exercises and practical examples, and with material or references explaining the background required to follow the course.

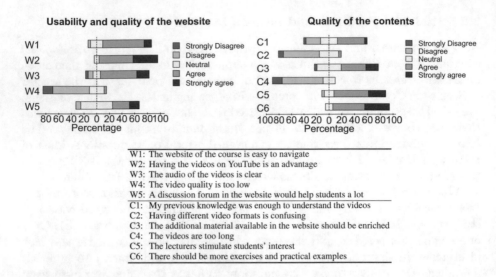

W1:	The website of the course is easy to navigate
W2:	Having the videos on YouTube is an advantage
W3:	The audio of the videos is clear
W4:	The video quality is too low
W5:	A discussion forum in the website would help students a lot
C1:	My previous knowledge was enough to understand the videos
C2:	Having different video formats is confusing
C3:	The additional material available in the website should be enriched
C4:	The videos are too long
C5:	The lecturers stimulate students' interest
C6:	There should be more exercises and practical examples

Fig. 2. Students' answers to the question "Express how much you agree to the following statements on the course *The art of grasping and manipulation in robotics*".

4   Conclusions and Future Work

In this work we presented a new MOOC on *The art of grasping and manipulation in robotics*, that is on-line since November 2017. Thanks to the preliminary feedback of 26 students, the course will be improved, and will be promoted within the robotics community starting from April 2018. The content will be continuously updated with new resources and cutting edge research topics.

Acknowledgement. We gratefully acknowledge the funding provided by "The Math-Works, Inc." for the development of the MOOC.

References

1. Amarnath, C.: Lecture Series on Robotics. Department of Mechanical Engineering, IIT Bombay (2008). www.youtube.com/watch?v=DaWMvEY3Qgc&list=PL2A735F42FA18D5DD
2. Bicchi, A., Kumar, V.: Robotic grasping and contact: a review. In: IEEE International Conference on Robotics and Automation, Proceedings, ICRA 20000, vol. 1, pp. 348–353 (2000)
3. Bongard, J.: Evolutionary robotics. University of Vermont (2016). https://www.youtube.com/watch?v=ANh_HIGmtRE&list=PLAuiGdPEdw0jySMqCxj2-BQ5QKM9ts8ik
4. Corke, P.: Robotics, Vision and Control. Springer Tracts in Advanced Robotics, vol. 73. Springer, Heidelberg (2011)

5. Corke, P., Greener, E., Philip, R.: An innovative educational change: massive open online courses in robotics and robotic vision. IEEE Robot. Autom. Mag. **23**(2), 81–89 (2016)
6. Coursera. Robotics specialization. University of Pennsylvania (2016). https://www.coursera.org/specializations/robotics#about
7. Cuevas, V.: Fundamentals of neuromechanics. University of Southern California (2016). http://valerolab.org/fundamentals/
8. de la Croix, J.P., Egerstedt, M.: Flipping the controls classroom around a MOOC. In: 2014 American Control Conference, pp. 2557–2562, June 2014
9. De Luca, A.: Robotics 1. Sapienza Università di Roma (2014). https://www.youtube.com/watch?v=pitZv3PuVMw&list=PLAQopGWllcyaq DBW1zSKx7lHfVcOmWSWt
10. Deimel, R., Brock, O.: A novel type of compliant and underactuated robotic hand for dexterous grasping. Int. J. Robot. Res. **35**, 161–185 (2015)
11. Egerstedt, M.: Control of mobile robots. Georgia Institute of Technology (2013). https://www.coursera.org/learn/mobile-robot
12. Fankhauser, P., Jud, D., Wermelinger, M.: Programming for robotics (ROS). Eidgenossische Technische Hochschule (ETH) Zurich (2017). https://www.youtube.com/watch?v=0BxVPCInS3M
13. Carbone, G. (ed.): Grasping in Robotics. Mechanisms and Machine Science, vol. 10. Springer, London (2013)
14. Khatib, O.: Introduction to robotics. Stanford Computer Science Department (2008). www.youtube.com/watch?v=0yD3uBshJB0&list=PL65CC0384A 1798ADF
15. Lynch, K.M., Park, F.C.: Modern Robotics: Mechanics, Planning, and Control. Cambridge University Press, Cambridge (2017)
16. Malvezzi, M., Gioioso, G., Salvietti, G., Prattichizzo, D.: SynGrasp: a matlab toolbox for underactuated and compliant hands **22**(4), 52–68 (2015). http://sirslab.dii.unisi.it/syngrasp/
17. Murray, R.M., Sastry, S.S., Li, Z.: A Mathematical Introduction to Robotic Manipulation, 1st edn. CRC Press Inc., Boca Raton (1994)
18. Prattichizzo, D., Trinkle, J.C.: Grasping. In: Siciliano, B., Khatib, O. (eds.) Springer Handbook of Robotics, pp. 955–988. Springer, Heidelberg (2016)
19. Reich, J.: Rebooting MOOC research. Science **347**(6217), 34–35 (2015)
20. Roa, M.A., Prattichizzo, D., Malvezzi, M., Pozzi, M.: Workshop on evaluation and benchmarking of underactuated and soft robotic hands (2016). http://benchsofthands-ws-iros16.diism.unisi.it/
21. Salvietti, G., Malvezzi, M., Prattichizzo, D., Brock, O.: Workshop on exploiting contact and dynamics in manipulation (2016). http://clem.dii.unisi.it/~malvezzi/wordpress/
22. Della Santina, C., Grioli, G., Catalano, M., Brando, A., Bicchi, A.: Dexterity augmentation on a synergistic hand: the Pisa/IIT SoftHand+. In: 2015 IEEE-RAS 15th International Conference on Humanoid Robots (Humanoids), pp. 497–503 (2015)
23. Siciliano, B., Sciavicco, L., Villani, L., Oriolo, G.: Robotics-Modelling, Planning and Control. Advanced Textbooks in Control and Signal Processing Series. Springer, London (2009)
24. Siciliano, B., Khatib, O.: Springer Handbook of Robotics. Springer, Heidelberg (2016)

25. Sun, Y., Berenson, D., Brock, O., Choi, H.R., Grupen, R.A., Jentoft, L., Messina, E., Moon, H., Roa, M.A., Santos, V.: Workshop on Robotic Hands, Grasping, and Manipulation (2015). http://www.rhgm.org/activities/workshopicra15/
26. Tedrake, R.: Underactuated robotics. Massachusetts Institute of Technology (2010). https://www.youtube.com/watch?v=Z8oMbOj9IWM&t=5s
27. Yuan, L., Powell, S.: Partnership model for entrepreneurial innovation in open online learning. E-learning Papers, 41 (2015)

Challenging Intensive Project-Based Education: Short-Term Class on Mobile Robotics with Mechatronic Elements

Anton Yudin[1]([✉]), Andrey Vlasov[1], Maria Salmina[2], and Vladimir Sukhotskiy[3]

[1] Bauman Moscow State Technical University, Moscow, Russia
skycluster@gmail.com
[2] Lomonosov Moscow State University, Moscow, Russia
[3] State Budget Vocational and Educational Institution "Vorobyovi Gori",
Moscow, Russia
http://www.bearobot.org

Abstract. The paper presents results of another successful short-term intensive technical education session carried out by the authors in the field of robotics with a group of high school students. The summary shows progression of steps to explain and realize technical issues in building a mobile robot, to test the obtained knowledge and skills and adds another arrangement step to the previous work.

Keywords: Educational robotics · STEM education
Intensive education · Digital fabrication
Human capability development

1 Introduction

The work described in this paper relates to after-school (optional) student activities in the field of engineering education. While school subjects provide general knowledge for students it is a common challenge to connect it to real-world application and practical use. The authors strive for this goal with STEM disciplines on the basis of educational robotics and digital fabrication.

Types of activities in educational robotics differ in form and content. Parameters which influence both include time (duration of an activity in time), available means (environment's specification and resources) and students' age.

Very short session (up to 10 h) aims at promotion of robotics and attraction of the youth to technical sciences and takes forms of a masterclass [1] or a workshop [2]. It can be divided in several parts presenting aspects of complex robotic systems or be a solid one-time experience.

Short session (up to 80 h) aims at gaining in-depth experience in one of the technical aspects of robotics for the purpose of further professional orientation and takes an intensive form (i.e. summer camp [3]) or an unintensive form (annual course) [4].

© Springer Nature Switzerland AG 2019
W. Lepuschitz et al. (Eds.): RiE 2018, AISC 829, pp. 79–84, 2019.
https://doi.org/10.1007/978-3-319-97085-1_8

General session (up to 200 h) aims at forming a competence in one of the technical aspects of robotics. It is meant to be an annual course and can be organized to be repeatedly taken by students for several years [5].

This paper provides another step in the authors's work towards a short session's structure and contents. Compared to the previous pedagogical experience this step includes parts to be used with unprepared, unexperienced (in terms of engineering and robotics) high-school students.

As many different scientific groups approach the educational robotics topic today it is worth mentioning that some of course share similar views on the contents of the educational process and means with the authors (for example [6–8]). Such work describes challenges of presenting and using today's and near-future technology to students with its depth to be realized in interesting and motivating form.

In this context it is worth mentioning that the proposed approach values technical creativity and ability to build original solutions to repeat as close as possible the work of professional engineers in real-life projects. Thus, educational tasks are presented in a form of competition rules, in a way that leads to diverse solutions.

While the presented session lacks the work on machines for students it still presents a custom toolbox produced with such digital fabrication equipment. The main aim being the promotion of educational robotics and the idea of its longer session's structure with a series of clean understandable steps for complete beginners.

2   Class Specifics

The 2-week long session was carried out in October 2017 with a group of high-school students gathered from different geographical regions selected upon their learning results in general school subjects and after-school activity. They were united in 6 teams of 5–7 members and given about 50 h of study and work time to finish a practical engineering project (Fig. 1).

While most of the audience didn't take robotics classes before the main idea for this session was to prepare a series of well-established parts to guide students' progress on the main challenge of building a remotely-controlled robot to win a competition at the end of the class (Fig. 2).

Figure 3 shows 3 basic solutions presented to students one after the other to form the educational progress starting from number 1 and finishing with number 3. Each of the 3 solutions has a main technical challenge to solve and when finished the effort of the team is accumulated to finally form a fully functional mobile robot.

The basic robots' parts were built with CNC machines of a common digital fabrication lab to show potential ways of using such technologies in engineering project work. When available such machinery can be easily added to the session with an idea of building custom robot parts.

The most simple and basic mobile robot shown as 1 in Fig. 3 was assembled in the first part of the educational session. The main challenge in this case was to build the right connection of the wire remote control's buttons, battery and the motors to move the robot chassis around the playing field.

In the beginning of this stage students had an example of the robot presented to them. The example was mainly a mechanical prototype lacking the needed wiring for correct operation. During the assembly process students had to "invent" it and test to understand the basics of robot control. The kit included premanufactured plastic parts, metal hardware, DC motors, wires, buttons, a relay module, a fuse and a battery. The equipment included multimeter, plastic bending machine, soldering station.

Before the assembly students were presented with the theory of motor operation directed at simple motion control of an electric motor. The student groups had to connect a single motor, buttons and a battery with wires to show they are able to do the reverse rotation of the motor's shaft.

After this was clear to everyone the challenge was to achieve the same results with a number of relays between each of the two motors and the buttons but this time implemented in a robot's chassis.

Figure 3 also shows the challenge of adding a radio module with a microcontroller (number 2) to substitute the wire in the previous solution and programm the behaviour of button controls.

The radio modules were given in a ready to use form. Each module had a number of contact pins available to transparently pass the signal from one end to the other. Students had to work out the connection diagram and assembly the improved control panel. The second radio module was to be included in the robot's control circuit to pass the information of the already well-established button presses to the motors.

Number 3 in Fig. 3 shows the challenge of adding a gripping unit and programming the controls to manipulate playing elements in the competition in precise manner.

The stage introduced servo motors and microcontroller programming of their control. Theory of operation and programming examples preceded hacking of the radio modules' microcontroller (Arduino compatible) on both ends to add grip capability to a robot.

To ease the start of such work students were presented with a well-commented program template in Arduino IDE to be changed for the desired robot behavior for each team. From now on students had to test their control on the field solving a number of tasks stated in the competition rules. The idea to solve more tasks in the given time stimulated optimization of such programming and motivated to actually interact with the playing field and its playing elements while operating the robots to find the winning strategy and train the robot pilots.

Fig. 1. Work-in-progress and educational environment.

Fig. 2. Competition educational environment and motivation.

Fig. 3. Project's progression steps for students.

3   Conclusion

The presented approach to project-based engineering education proved to be a good practice for short-term class organization with a very good potential for tuning the class' contents according to the conditions, aims and time.

Happen it the involved students decide to continue in the robotics field they definitely have a solid example of what they can achieve from scratch using the digital fabrication equipment available in labs around the world. It is especially important for the university ones, as most of the students are near the enroll age and wish to enter universities soon. The solution presented to them is basic in its core, without any types of black boxes. Thus as a configurable blueprint it can be used in their future project work for more complex designs [9,10].

While basic parts of the session stay same to provide the novice students with bearable basic experience of engineering work, the educational environment can be adjusted to the real conditions of the venue and student knowledge: CNC machines and materials availability, prior programming and engineer design experience. Such adjustments influence difficulty of tasks, time to finish, level of independent (self-dependent) behaviour and creativity.

With adjustments in mind the presented short-term class core can be widely used for promotion of engineering education among youth, provision of the first experience of engineering project-based practice for professional orientation or even for intensive educational sprints for experienced students.

References

1. Yudin, A., Sukhotskiy, D., Salmina, M.: Practical mechatronics: training for mobile robot competition. In: 6th International Conference on Robotics in Education. RiE 2015, pp. 94–99 (2016). ISBN 978-2-9700629-5-0
2. Yudin, A., Salmina, M., Sukhotskiy, V., Dessimoz, J.-D.: Mechatronics practice in education step by step, workshop on mobile robotics. In: 47th International Symposium on Robotics. ISR 2016, pp. 590–597 (2016)
3. Yudin, A., Vozhdaev, A., Sukhotskiy, D., Salmina, M., Sukhotskaya, T., Sukhot-skiy, V.: Intensive robotics education approach in the form of a summer camp. Adv. Intell. Syst. Comput. **560**, 246–250 (2017)
4. Sukhotskiy, D., Yudin, A.: Startup robotics course for elementary school. In: Communications in Computer and Information Science, CCIS, vol. 156, pp. 141–148 (2011).https://doi.org/10.1007/978-3-642-27272-1_13
5. Salmina, M., Kuznetsov, V., Poduraev, Y., Yudin, A., Vlasov, A., Sukhotskiy, V., Tsibulin, Y.: Continuous engineering education based on mechatronics and digital fabrication. In: 6th International Conference on Robotics in Education. RiE 2015, pp. 56–57 (2016). ISBN 978-2-9700629-5-0
6. Hirschmanner, M., Lammer, L., Vincze, M.: Mattie robot - a white-box approach for introducing children with different interests to robotics. In: 6th International Conference on Robotics in Education. RiE 2015, pp. 109–111 (2016). ISBN 978-2-9700629-5-0
7. Lopez-Cazorla, D., Cardenas, M.-I., Garcia-Rodriguez, M., Tarradellas-Vinas, O.: Starting robotics in secondary school: assembling i-SIS an autonomous vehicle. In: 6th International Conference on Robotics in Education. RiE 2015, pp. 132–137 (2016). ISBN 978-2-9700629-5-0
8. Agatolio, F., Moro, M.: A workshop to promote Arduino-based robots as wide spectrum learning support tools. Adv. Intell. Syst. Comput. **457**, 113–125 (2017). https://doi.org/10.1007/978-3-319-42975-5_11
9. Yudin, A., Kolesnikov, M., Vlasov, A., Salmina, M.: Project oriented approach in educational robotics: from robotic competition to practical appliance. Adv. Intell. Syst. Comput. **457**, 83–94 (2017). https://doi.org/10.1007/978-3-319-42975-5_8
10. Vlasov, A., Yudin, A.: Distributed control system in mobile robot application: general approach, realization and usage. In: Communications in Computer and Information Science, CCIS, vol. 156, pp. 180–192 (2011). https://doi.org/10.1007/978-3-642-27272-1_16

Using Finite State Machines
in Introductory Robotics

Richard Balogh[1]([⊠]) and David Obdržálek[2]

[1] Slovak University of Technology in Bratislava, Bratislava, Slovakia
richard.balogh@stuba.sk
[2] Charles University, Prague, Czech Republic
David.Obdrzalek@mff.cuni.cz

Abstract. In this paper, we describe the use of the finite state machines (FSMs) in introductory robotics, especially for implementing entry-level control systems. The paper aims to show FSM can be a good tool for implementing control system of a robot, powerful as well as easy to be used. Although it may seem as overkill for typical first exercises with robots, their knowledge can significantly help when the task complexity grows.

Keywords: Finite state automata · Educational robotics
Robotic competitions

1 Introduction

Finite state machines (FSM) are one of the many types of automata. In their hierarchy, they lie on a low level, i.e. their relative power is not big in comparison with other types (e.g. Push-down automata or Turing machines). However, they are well sufficient for many tasks in various applications and areas, including robotics. Moreover, they can contribute to implementation readability and maintainability as they represent a methodical way of using simple construction parts for forming a powerful system. Therefore we believe FSMs are of a great use also for introductory robotics, especially in education. However, a very typical development process in such cases is creation of a simple dedicated system which is sufficient for solving the first few tasks and then gets improved over time to cover more functionality and resolve special cases and specific details. This often leads to a system which grows well beyond understandability and maintainability because of its nature of a monolith system with numerous patches and exceptions. In our paper, we would like to motivate the readers who were not aware about the usability of FSMs in introductory robotics to use them and to show that using FSMs and their enhanced versions is not difficult at all and can bring the aforementioned advantages which in final effect helps the users to focus more on problem solving than on its low level implementation.

© Springer Nature Switzerland AG 2019
W. Lepuschitz et al. (Eds.): RiE 2018, AISC 829, pp. 85–91, 2019.
https://doi.org/10.1007/978-3-319-97085-1_9

The rest of this paper is structured as follows: In Sect. 2, the Finite State Machine is briefly introduced. Section 3 shows two examples of FSM used for implementation of simple automata, and Sect. 4 concludes the paper by summarizing the arguments for using FSMs even by beginners.

2   Finite State Machines

In mathematics, automata theory is a relatively young topic. It was first studied in the second quarter of 20th century with a need for finite description of infinite objects in logic. Although it was originally a purely theoretical tool, it soon relieved it has a wide practical usage and became to be used in real-life to build control systems. For example, a special subset of FSM is a sequential function chart (SFC) which is used as a graphical programming language for programmable logic controllers (PLCs). It is one of the five PLC languages defined by the international standard IEC 61131-3 [1] and is widely used in industry.

Figure 1 shows an example of a simple FSM featuring three states S_1, S_2, S_3, and four transitions T_{1a}, T_{1b}, T_2, T_3. On the left, it is represented as a directed graph where nodes represent states and edges represent transitions. When a transition triggers because of the input associated with it (represented as edge value), the state machine changes its actual state and makes a transition to another state. The table in Fig. 1 (middle) shows a transition table where the current state is the row index, input is the column index and the resulting state is at their intersection. Figure 1 (right) shows a FSM tree where the start is in the root node and the tree expands in accordance to the transitions; whenever a state already existing in the tree is reached, it is not further expanded.

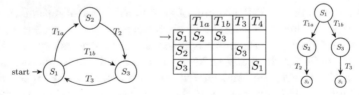

Fig. 1. General example of a state machine represented as a graph, table, and tree.

There are many possible implementations. In this paper, we show some basic ones using simple native constructs of common programming languages, suitable for beginners. Of course, it is also possible to implement the FSM in advanced ways and also as part of more complex systems. One of the possible implementations of the transition function in the C programming language is shown[1] in Listing 1.

---

[1] Throughout this paper, only relevant parts of the code are presented; the code snippets are not fully compilable and should serve as illustrations of the principles.

```
enum states {S1, S2, S3};
enum states currentState = S1;

switch( currentState ) {
  case S1: if( T1a() ) currentState = S2;
           else if( T1b() ) currentState = S3;
           break;
  case S2: if( T2() ) currentState = S3;
           break;
  case S3: if( T3() ) currentState = S1;
           break;
}
```

Listing 1. Transition function example.

Here, the inputs are implemented as calls to respective functions, returning boolean values corresponding to the input checks.

In typical implementation, states are represented as enumerations, inputs as functions providing the input read or checking the respective input appearance, and outputs as procedures performing the required action.

On lower scale, general advice for direct FSM implementation is to split the code into two parts – one with the transitions only and one with the execution of outputs. The FSM transitional part may be formed by a set of checks for possible transition triggering input(s) for each individual state. It could be performed either for each state evaluating the inputs or their combination to select the transition to a new state (as e.g. in Listing 1), or the other way round evaluating the inputs (or their combination) first and then selecting the transition to a new state based on the original state. Both options are of the same power and the selection typically depends on the density level of the state × input product. Secondly, the FSM executional part may be formed by nested `if` statements (or compound statements like `switch...case` in C, C++, or `case...when` in Ruby), or indexing a table containing the state-dependent functionality (as most states imply some action so such table would not be sparse).

If the state machine is not a simplest one, more typically a general FSM framework is created (or reused) and the programmer provides data for the machine, typically consisting of a table where each row represents all necessary data for one state, at least state activity function and list of new states based on individual inputs used as column index (see Listing 3 or [2]).

3   Examples

3.1   Line Following Robot

In Fig. 2, we show a basic line following robot control as this task is a typical first exercise in robotics. In this example, the robot is very simple and is equipped with only 2 line sensors working as follows. When a sensor detects the line (black), it outputs logical 1, otherwise it outputs 0. This task is typically solved as a reactive system where the output for the motor controllers is directly linked with the reading of the two sensors. However, we will show an implementation using a FSM which is better scalable.

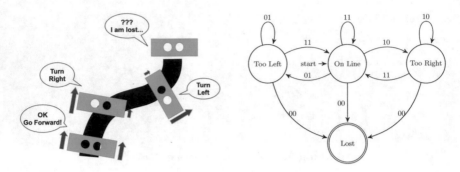

Fig. 2. Line following robot. Sensors detecting line at different phases (left) and corresponding FSM diagram (right).

With two binary sensors, we have in total four different states. Controlling two wheels, there are also four principally different output actions. For simplicity, we don't consider any activity after the robot is lost. For a code sketch, see Listing 2. As mentioned earlier, this could be implemented also as a reactive system, not dependent on the current state. In the listing, this "reactiveness" results in the state table being very similar for most of the states.

Although using a FSM for the line following task might look superfluous and overkilling, it is very easy to understand and can be easily further enhanced: for example, one might want to solve the situation when the robot gets lost, add more sensors, allow the line to split and join or contain crossings, extend the environment by additional marks on the ground etc. Adapting for such extensions means in case of FSMs adding more states, properly connect them using the transitions, and associating sensor reading and action functions, but does not need changing the already existing FSM core (e.g. see Listing 3: adding new state would mean writing the state action function, adding one line in the FSM array, and change some of the transitions in the remaining lines). This is very good for beginners.

```
// FSM states and transition table definitions:
  enum states {Lost, TooLeft, TooRight, OnLine};
  enum states stateTable[4][4]={
    {Lost, Lost, Lost, Lost},
    {Lost, TooLeft, TooRight, OnLine},
    {Lost, TooLeft, TooRight, OnLine},
    {Lost, TooLeft, TooRight, OnLine} };
// Transition:
  transition = readSensors();
  nextState = stateTable[currentState][transition];
// Action:
  const int slow=25, fast=75;
  switch(currentState) {
    case Lost:
      leftMotor.stop();
      rightMotor.stop();
      break;
    case TooLeft:
      leftMotor.run(fast);
      rightMotor.run(slow);
      break;
```

```
      case TooRight:
        leftMotor.run(slow);
        rightMotor.run(fast);
        break;
      case OnLine:
        leftMotor.run(fast);
        rightMotor.run(fast);
        break;
    }
```

Listing 2. Line following robot state machine - direct implementation

3.2   Ketchup Cans Collecting Robot

The FSM has been successfully used also for implementing the core control of the "MART Friday Bot". This robot was originally prepared for a choreography-based "Robot Dance" performance but thanks to using a FSM, it was successfully changed to the object manipulation task of the "Ketchup House" competition [3] at Istrobot event in just a few hours with sound results[2]. As relieved from talks to other robot builders in this competition, their software implementations differed most notably in lack of global code structure and contained typically lots of code functionalities merged into a single big program as a result of continuous improvements. That was typically hard to debug and close to impossible to adapt. On the other hand, the MART Friday Bot code was composed of simpler independent functions which served the FSM in performing individual actions. The change in the robot task did not affect those functions neither the FSM core, only the definitions of some states, transitions, and inputs triggering the transitions, closely following the principles shown in Listing 3.

```
// FSM core definitions:
  enum states {Lost, TooLeft, TooRight, OnLine};
  typedef int (*actionFunc)(enum states,int);
  typedef struct {actionFunc action; enum states next[4];} stateFunction;
// FSM state action and transition definition:
  stateFunction FSM[4] = {
    {handleLost, {Lost, Lost, Lost, Lost}},
    {handleTooLeft, {Lost, TooLeft, TooRight, OnLine}},
    {handleTooRight, {Lost, TooLeft, TooRight, OnLine}},
    {handleOnLine, {Lost, TooLeft, TooRight, OnLine}} };
// Transition:
  transition = readSensors();
  nextState = FSM[currentState].next[transition];
  currentState = nextState;
// Action:
  FSM[currentState].action(currentState,transition);
// Action procedures:
  const int slow=25, fast=75;
  int handleTooRight(enum states state,int input) {
    leftMotor.run(slow);
    rightMotor.run(fast);
  }
// etc. for all other action procedures
```

Listing 3. Line following robot state machine - scalable implementation

---

[2] The robot placed third in 2016 and second in 2017 editions of the Istrobot competition.

4   Conclusions

In this paper, we have presented the Finite State Automata as a possible tool for implementing the core of a robot control. We have shown several examples which should give the reader basic understanding of FSM usage in this area. Our experiences show there are numerous task types where a FSM can significantly help at the implementation phase even if the task seems to be very simple at the first glance. Using the FSM can make the controlling code be readable, understandable and maintainable which is very important especially in education.

Although there exist many implementations, libraries or systems based on FSMs, many of them are way too complex for beginners to understand and cannot be recommended as a starting point. Instead, we have shown how a simple FSM can be programmed from scratch in a simple way. That helps to understand the principles and allows for a fast progress towards real world usage.

Reasons to Use FSMs in Robotics Education

- Finite State Automaton it is a powerful tool for describing sequential behavior of a control program and is easy to understand.
- Using the FSM helps to partition a control problem into a smaller units while at the same time it provides clear and global overview of the process.
- If the program is to be modified, usually it is done at single point only. Modifications to the control section do not affect settings of the outputs.
- The program is easy to test. For example, if it gets stuck, it is easy to determine which transition is not working or which condition is missing.
- There exist simulators and audit tools (e.g. MathWorks Stateflow [4]) so it is possible to test the state machine even before obtaining the real hardware.
- It forces an user to structure the programs which in turn minimizes the chance of programming errors and improves documentation.
- Learning curve for using FSM is minor. It is likely the reader already knows about FSMs from math or logic courses; even if not, it is easy to understand.

On contrary, it should be also noted that FSM is not a universal magic tool capable of solving everything. There are many areas where using FSM is not appropriate or where it can not be used at all. If nothing else, one should be always aware of its theoretical properties: taking into consideration the Automata theory and Formal grammars, the FSM is able to handle only regular languages which might not be sufficient for solving real world tasks.

Full version of this text is available online at [5].

Acknowledgements. This paper was partially supported by the grants of the Research Agency of the Ministry of Education, Science, Research and Sport of the Slovak Republic (VEGA) 1/0819/17 Intelligent mechatronics systems (IMSYS).

References

1. IEC 61131-3 International Standard. Programmable controllers – Part 3: Programming languages. International Electrotechnical Commission, Geneva, Switzerland (2013). ISBN 978-2-83220-661-4
2. Amlendra: How to implement finite state machine in C (2017). https://aticleworld.com/state-machine-using-c/
3. Balogh, R.: Ketchup house – a promising robotic contest. In: 3rd Conference on Robotics in Education 2012. Matfyzpress, Prague, Czech Republic, vol. 3, pp. 41–45 (2012). ISBN 978-80-7378-219-1
4. MathWorks: Stateflow – Model and simulate decision logic using state machines and flow charts (2018). https://www.mathworks.com/products/stateflow.html
5. Balogh, R., Obdržálek, D.: Using finite state machines in introductory robotics (full text) (2018). http://robotics.sk/go/FSM

Teacher Training in Educational Robotics

An Experience in Southern Switzerland: The PReSO Project

Lucio Negrini[✉][iD]

Department of Education and Learning, University of Applied, Sciences and Arts of Southern Switzerland (SUPSI), Piazza San Francesco 19, 6600 Locarno, Switzerland
Lucio.negrini@supsi.ch

Abstract. The PReSO project is a pilot project with 17 teachers of the pre-primary and primary schools in Southern Switzerland. Aim of the project is to introduce children to computational thinking and to foster their interest towards ICT and the STEM disciplines through educational robotics. However, it is challenging to introduce educational robotics into compulsory schools since there is for example a lack of didactic materials and teachers often perceive robotics as something too difficult. To bypass these obstacles and introduce robotics into schools, the research team has developed a teacher-training concept in educational robotics and has trained 17 teachers. The project has achieved good results indicated by the fact that most of the participating teachers have integrated robotics into their annual program. This article describes the training concept and discuss how to train compulsory school teachers in the field of educational robotics.

Keywords: Teacher training · Educational robotics · Computational thinking

1 Introduction

Digitization and automation have entered our everyday lives and have brought a profound structural change in the working world [1]. Today we can find robots that manage the logistics, others that assemble pieces and build various objects from pens to cars, some distribute drinks and snacks, and others greet you at the hotel reception.

Because of these changes, new work processes and new job profiles have emerged. A factory worker for example will no longer need to assemble the pieces of a car but will have to control or program the robot that assembles the pieces; a journalist will have to be able to manage content on social media and a mechanic will have to know how to reprogram the control unit of the cars. These new processes require new skills and competences. The Swiss federal report on digital education [2] states for example that active citizenship today is not imaginable without basic digital and media literacy. In addition, the World Economic Forum [3] lists as key competences of the 21st century the ICT competences as well as transversal competences such as critical thinking, creativity, communication or cooperation. Furthermore, STEM skills

© Springer Nature Switzerland AG 2019
W. Lepuschitz et al. (Eds.): RiE 2018, AISC 829, pp. 92–97, 2019.
https://doi.org/10.1007/978-3-319-97085-1_10

(science, technology, engineering, and mathematics) are compelling for an increasing number of jobs and should be required in schools and colleges.

How can these skills and competences be promoted at school and thus prepare our students for the 21st century society? A possible answer is through educational robotics. In fact, educational robotics allows introducing the basic concepts of computational thinking, to learn how technologies work and to gather ICT and programming competences as well as transversal competences.

To introduce educational robotics into school, teachers must be trained. In this article, we discuss how to train teachers in the field of educational robotics, describing an experience carried out in southern Switzerland in the Canton of Ticino.

2 Framework

Compulsory school in Ticino consists of three school orders: pre-primary school (3 years, pupils from 3 to 5 years of age; the first year is not compulsory), primary school (5 years, pupils from 6 to 10 years of age) and the lower secondary school (4 years, pupils from 11 to 14 years of age).

Media, technology and educational robotics have only a tiny space in the curricula of compulsory school in Ticino [4]. In the curricula of the pre-primary school and of the first two years primary school, educational robotics is not explicitly cited. However, the curriculum of the last three years of primary school foresees that children should learn how to program technological objects. In the curriculum of the lower secondary school, computational thinking and robotics are also present and there exists an optional course in educational robotics for the fourth (last) class of the lower secondary school. In all three curricula, activities in this field are classified as general training, i.e., skills that are not part of one or more specific disciplines but involve all disciplines, and are therefore mostly done in form of school projects that last only few days.

Although the school curricula mention robotics, only few teachers in Ticino carry out such activities. The problem of the introduction of educational robotics into schools is an open one [5–7]. Possible reasons for this absence can be a lack of didactic materials [6], fear of lacking computer science skills [8] or few knowledge of the educational benefits of robotics [9].

To increase the presence of educational robotics into the schools in Ticino we have started at the SUPSI-DFA a project called PReSO (progetto robotica e scuola dell'obbligo).

3 The "PReSO" Project

The PReSO-project is an action research project [10] that has been carried out in the school years 2015/16 and 2016/17 in Ticino.

The aim of the project was to introduce children to computational thinking and to foster their interest towards ICT and the STEM disciplines using digital didactic tools (e.g. educational robots). To reach this objective we decided to develop a teacher-training concept in educational robotics. We recruited five pre-primary teachers (all female) and 12 primary school teachers (5 male, 7 female), who did not have specific knowledge in the field of information technology or computer sciences, and trained them during 12 half-days distributed over two years.

During the first year, the participating teachers followed a training that included an introduction to educational robotics (history of the educational robotics, the different approaches of the educational robotics, the competences that are developed with educational robotics and ethics) the concept of computational thinking [11, 12], learning theories and teaching methods like the approach of the constructionism of Papert [13], and the project method [14], and technical knowledge about the proposed educational robots. In a second phase, the trained teachers introduced the educational robots in their classes and tested the didactic activities they have designed. In the second year of the project, the teachers defined a long-term project integrating the robots, in order to pursue various objectives, including the development of computational thinking. During the two years of the PReSO project, we worked closely together with the teachers, experimenting and developing with them various didactic activities of educational robotics.

3.1 Educational Robotics

In the PReSO project, we have mainly used two educational robots: Blue-Bot and Thymio II. Teachers of the pre-primary schools and of the first two years of primary school used Blue-Bot. Blue-Bot enables children to approach the basics of programming by inventing their first algorithms to move the robot forward/backward and left/right. For children of the last three years of primary school we used the Thymio II robot created by the Lausanne Polytechnic and programmable with a visual programming language as well with a text programming language. Thymio II features a wide range of sensors (nine IR proximity sensors, a three-axis accelerometer, a microphone, a temperature sensor, a remote control receiver, an SD-card slot, and five capacitive buttons) as well as two motors, a loud-speaker, and 39 LEDs spread all over its body [8]. A sensorized robot like Thymio II greatly expands programming activities and allows to introduce a reflection on the sensor-actuator loop and on the event-driven programming. Thymio also allows to modify gradually the difficulty through a selection of several programming languages and environments: from the basic behaviours of Thymio that are accessible also to young children, to the visual programming language until the text programming. This feature of Thymio was appreciated by the teachers, that could develop activities accordingly to the competences of their pupils. Blue-Bot, however, is more limited to a sequential programming and the possible activities run out quickly.

It was important to let the teachers try to use these robots and explain them how they work (for example how sensors work, what is an actuator and so on). We organized two session were we presented the robots and made some activities where teachers could explore all the functionalities of the robots and get familiar with them. This allowed to see the robots with less fear and to notice that there are not requested high technological skills to use them in compulsory school for basic activities. If teachers wanted to plan more challenging situations where more programming skills were requested, than they involved also the research team as experts.

3.2 Didactic Activities

One central element in order to introduce robotics into schools are the didactic activities. Teachers need materials and good examples that they can use in schools. In order to create a database of educational robotics activities, we asked all teachers implicated in the PReSO project to develop some activities and to test them into their school classes. During this work, we guided the teachers and reviewed their activities. The teachers have produced many materials that they exchanged among each other. All the materials have been tested in the classes and ameliorated when needed.

The introduction of educational robotics to the pupils has mainly been made through the lecture of a book about robots, followed by a discussion with the pupils about their conceptions of robots. Questions like: "what is a robot?" "how does a robot works?" have been discussed with the pupils. After this first phase a brief introduction about the programming languages has been made. How can we program a robot? Children have developed their own programming languages and tested them, becoming themselves robots. In this case a child "programs" one of his classmates. First exercises with the robots have been introduced, learning for example to build a sequence or to make a loop with Blue-Bot or the conditions and the sensor-actuator loop with Thymio II. After this training phase, during the second year of PReSO, teachers and children together have developed in their classes some long-term robotics project following the approach of the project method. So for example a teacher have built with the children a game for the robots, another one have putted on stage "the eye of the wolf" by Pennac using the Thymio II robots as the main characters of the story and another one have created the Olympics games for robots. The research team was available in case of need.

3.3 Evaluation

The PReSO project has achieved good results indicated by the fact that most of the participating teachers have integrated robotics into their annual program. In addition, the research team and the trained teachers have proposed educational robotics activities to other schools and to the public as for example during the events Matematicando, Asconosc(i)enza[1] or FIRST LEGO League[2], starting a discussion on educational robotics in compulsory school also in Ticino.

The chosen training concept was successful. Teachers could approach robotics without fear and have noticed how robots can be used into classes and which educational benefits they can bring. Working together with teachers on their activities allowed creating a database of materials that can be used also by others teachers that want to introduce robotics into school but do not know how. Furthermore, it is important to consider also the teaching methods that teachers use. Bringing a robot into the classroom is not enough to build skills. The project method seems to be a good method for carrying out educational robotics activities. Pupils can work together on different projects and build also transversal competences such as collaboration,

---

[1] Matematicando and Asconosc(i)enza are two events to make science accessible to the general public.

[2] First Lego League is a robotics tournament.

communication or creativity. Working together with teachers also allowed to build a community and to remain in contact with the teachers. The research team is regularly invited in the classes to see the activities that teachers have planned or to assist them in some activities. Some classes also come and do the robotics activities together with the research team in the spaces of the SUPSI-DFA where they can use the technological infrastructure that is not always present in their classes. In doing so, we have an overview on what teachers do and what they need also after the project has official ended.

4   Conclusion and Future Work

As we know from the literature, it is challenging to introduce educational robotics into compulsory schools since there is for example a lack of didactic materials and teachers often perceive robotics as something too difficult. To bypass these obstacles and introduce robotics into schools, the PReSO team has developed a teacher-training concept and has trained 17 teachers. The training included an introduction to educational robotics, the concept of computational thinking, the project method, technical knowledge about the proposed educational robots and the development of didactic activities.

In order to reach the teachers and gain their interest it was important to show them which benefits robotics can bring. We decided to introduce the concept of computational thinking and show how this concept is important for the 21st century society. It was also important to accost this concept to their everyday life and to show that computational thinking and programming are not only skills needed by computer scientist and that working on this concept can bring benefits for everybody.

Teachers often have the belief that they need high technological competences in order to program a robot. In this case, it was important to start with easy programming exercise that everybody could solve and to increase gradually the difficulty. At the end, it was for all clear that it is not necessary to have high technological competences to program an educational robot for basic activities.

Another element that teachers appreciated was the accompaniment of the research team during the development of didactic activities with the robots and also after the formation. Teachers needed to be reassured that their activities worked well. In addition, this allowed creating a database of didactic activities that all teachers could use and a community between teachers and experts.

Since PReSO was a pilot project and only few teachers were trained, the research team decided to develop a more structured training in order to reach more teachers. Since September 2017, SUPSI-DFA in collaboration with the Department of Innovative Technologies of SUPSI (DTI), offers therefore a Certificate of advanced studies (CAS) in educational robotics.

The CAS has started with 22 teachers from primary and lower secondary schools, half of them are female.

For the future, we would like to reach more teachers and to constitute a community and a database of robotics activities that can be done in compulsory school.

References

1. Genner, S.: Digitale Transformation: Auswirkungen auf Kinder und Jugendliche in der Schweiz – Ausbildung, Bildung, Arbeit, Freizeit. ZHAW Zürcher Hochschule für Angewandte Wissenschaften, Zürich (2017)
2. Schweizerische Eidgenossenschaft: Bericht über die zentralen Rahmenbedingungen für die digitale Wirtschaft. Schweizerische Eidgenossenschaft, Bern (2017)
3. World Economic Forum: New Vision for Education: Fostering Social and Emotional Learning through Technology. WEF, Geneva (2016)
4. Dipartimento Educazione Cultura e Sport: Piano di studio della scuola dell'obbligo ti-cinese. Dipartimento Educazione Cultura e Sport, Repubblica e Cantone Ticino, Bellinzona (2015)
5. Benitti, F.B.V.: Exploring the educational potential of robotics in schools: a systematic review. Comput. Educ. **58**(3), 978–988 (2012)
6. Mubin, O., Stevens, C.J., Shahid, S., Mahmud, A.A., Dong, J.-J.: A review of the applicability of robots in education. Technol. Educ. Learn. **1**(1), 1–7 (2013)
7. Riedo, F.: Thymio: a holistic approach to designing accessible educational robots. Ph.D. thesis, EPFL, Lausanne (2015)
8. Chevalier, M., Riedo, F., Mondada, F.: Uses of Thymio II: how do teachers perceive educational robots in formal education? IEEE Robot. Autom. Mag. **23**(2), 16–23 (2016)
9. Alimisis, D.: Educational robotics: open questions and new challenges. Themes Sci. Technol. Educ. **6**(1), 63–71 (2013)
10. Elliott, J.: Action Research for Educational Change. Open University Press, Buckingham (1991)
11. Barr, V., Stephenson, C.: Bringing computational thinking to K-12: what is involved and what is the role of the computer science education community? ACM Inroads **2**(1), 48–54 (2011)
12. Wing, J.: Computational thinking. Commun. ACM **49**(3), 33–35 (2006)
13. Harel, I., Papert, S. (eds.): Constructionism. Ablex Publishing Corporation, Westport (1991)
14. Kilpatrick, W.H.: The Project Method: The Use of the Purposeful Act in the Educative Process. Teachers College, Columbia University, New York City (1918)

Impact Evaluation

Bringing Educational Robotics into the Classroom

Implications of a Robotics Promotion Program

Benedikt Breuch and Martin Fislake[✉]

Universität Koblenz-Landau, Campus Koblenz, Universitätsstraße 1,
56070 Koblenz-Metternich, Germany
{bbreuch, fislake}@uni-koblenz.de

Abstract. In this paper we present the findings of our survey that deals with the effects of a program to promote robotics in education in Rhineland-Palatinate, Germany. The state sponsors participating schools with LEGO® NXT/EV3 robot sets and the entry fee for the robotics contest FIRST® LEGO® League. In return, participating schools have to partake in the contest and develop a sustainable concept for the implementation of robotics in education. We briefly describe the program as well as the contest and, afterwards, present the research results followed by a discussion of the findings, showing that this type of promotion program is an effective tool to bring robotics into the classroom.

Keywords: Robotics · Education · Survey · Effects · Germany
Ministry of education · FIRST LEGO league · Promotion program
STEM · MINT

1 Introduction

In 2010, the Ministry of Education of Rhineland-Palatinate, Germany, initiated the so-called project "Robotics" to promote the interests of students in STEM. The project is part of the state program *Medienkompetenz macht Schule*[1]. The project aims at fostering the use of robots in class as well as in working groups. To achieve these objectives the ministry has been sponsoring each of 108 participating schools in the state of Rhineland-Palatinate with two LEGO® NXT/EV3 robot sets, the entry fee for the FIRST® LEGO® League (FLL), an FLL playing field, and a workshop in a project school [1].

Additionally, all schools need to submit a sustainable concept for the implementation of robotics in education. It is also part of the project that all participating schools partake in the FLL, a promotional program that is targeted at getting young people enthusiastic about science and technology and imparting team spirit.

Our survey determines what happened with the sponsored robots, the requisite educational concepts that deals with the application of robots in education, what effects

---

[1] It is the name of a state program to enhance media competence among students.

© Springer Nature Switzerland AG 2019
W. Lepuschitz et al. (Eds.): RiE 2018, AISC 829, pp. 101–112, 2019.
https://doi.org/10.1007/978-3-319-97085-1_11

may be reported by the involved schools up to now, and what experiences the teachers had during and after the project and their participation in the FLL.

2 Investigation of the Field

The state's promotion program follows the substantial literature in the wider field of educational robotics where the positive effects of robotics in educational environments are reported.

In this context, Eguchi [2] stresses the fundamental importance of approaches and tools that bring educational robotics into formal educational classroom settings so that such learning experiences are accessible to all students, not only to those who are privileged or boys.

Thus, educational robotics have been observed as an effective tool to change and support STEM education [3] as well as for facilitating students' STEM learning in a wide range of different skills [4, 5].

Furthermore, the project follows the implications of Nugent et al. [6] and Karp and Maloney [7] who recommend robotics challenges as a successful tool that "increases participants' performance in STEM disciplines and their likelihood to choose a career in this field".

3 The Project "Robotics"

According to the Ministry of Education's aim to promote the interest of students in STEM, 102 secondary and 6 vocational schools in Rhineland-Palatinate received two LEGO® robot sets, the entry fee for the FLL, an FLL playing field, and a workshop in a project school since the beginning of the project in 2010. Furthermore, additional sponsors sponsored one additional set for each of 60 schools until 2012 [1]. At the beginning, the sponsored robot sets consisted of the then current model LEGO® NXT and now of the latest model, the LEGO® EV3.

To achieve the ministry's aims participating schools have to submit a sustainable concept for the implementation of robotics in the classroom, although, there is no specification of any criteria for these concepts mentioned [1]. Since the beginning, project schools have to partake at least once in the FLL which claims to foster young people's curiosity in science and technology [4].

But, the FLL is not only about science and technology. According to the tournament information, it is a promotion program that wants to arouse interest in engineering and IT professions but also aims at promoting team spirit among young people [9].

In annual competitions, the participants solve complex tasks connected to current issues of science and technology, for instance, "Energy Crisis, Climate Change, Food Security, [and] Natural Disasters." [9].

In case of current budget limitations, the ministry decided to reduce the possible participants to a smaller number. Since the costs add up to approximately 1,350 € per school, only one district of the educational authorities within the state (they all have different sizes and numbers of schools) and, hence, only a few schools, are allowed to partake per year [1].

4   The Survey

4.1   Research Design

In order to study the effects of the ministry's robotics project, we conducted a study among 669 secondary and vocational schools. Schools that already participate in the ministry's project received an e-mail invitation for answering an online questionnaire that asks questions about the school's experiences with the project and what effects it had with respect to the use of robots in education. For the purpose of determining differences between now and the initial situation, the survey participants were also asked to give information about the initial robot and computer inventory and about the use of these robots before the school entered the project.

Schools that are not part of the project were asked to answer a different online questionnaire that contains questions about the current use of robots and the reasons for the school not to participate in the program. Both participants and non-participants were asked to assess the learning effects of robotics in education with respect to social competency, professional competence, media competence, and self-competence.

People answered the questionnaires anonymously and personal data other than age and sex was not queried. A subsequent assignment of the data to the schools was not intended. The research period started on January 29th, 2018 and ended on February 11th, 2018.

4.2   Research Results

Response Rate. The response rate of schools that participate in the project "Robotics" is 40%. Almost half of the responding schools are *Gymnasien* with a proportion of 48%, 32% *Realschulen plus*, 11% are integrated comprehensive schools, 7% are vocational schools, and 2% are schools with focus on learning and mental development.

In contrast to the participating group, the percentage of *Gymnasien* that are not part of the project and answered the question about their school type is 17% and, therefore, significantly lower. 30% of the responding non-participants who told us their type of school are *Realschulen plus*, while the share of integrated comprehensive schools of this group is 11%. The proportion of vocational schools among the responding non-participants is at 27% almost four times greater than the share of the participating vocational schools. 16% of the responding nonparticipants that revealed their school type are schools with focus on learning and mental development. Most strikingly, however, the proportion of *Realschulen plus* among the responding schools of both groups is almost the same. Taking the response rates of the two groups in this complete survey into consideration, representativeness for both the participating and non-participating schools can be assumed.

Contest Participation and Reasons. Of the schools that answered the question about their participation in the FIRST® LEGO® League, 9.1% answered that they are new to the project and that they partake in the FLL for the first time. 20% answered that they are about to participate in the FLL for the second time. A majority of 71.4% partook in the FLL at least two times or more.

First-time participants were asked about their expectations with regard to the project "Robotics" and the FLL. These questioned people state that, by partaking in the FLL, they want to arouse the student's interest in computer science and that they see this as an opportunity to initiate a robotics workgroup in their school. 56.3% of the schools that already are part of the project and took part at least once in the contest want to participate again in the FLL.

The most frequently mentioned reasons for a repeated participation where the students' motivation and wish to participate (72.2%) and positive experiences made in former contest (66.7%). With 28.1%, less than a third are sure that they do not want to partake in the contest again. The most frequently named argument for not partaking in the FLL again is the efforts exceeding the benefits (44.4%). 22.2% named missing motivation of the students and another 22.2% named costs as reasons to refrain from another participation in the contest.

Non-participation in the Project and Reasons. Schools that are not part of the project "Robotics" were asked for the reasons for their non-participation. Of the schools who answered the question, 50.0% do not know that the project exists. 13.2% are not interested in the project and 1.3% are interested but cannot enter the project because it is not being offered in their school district at this time. 34% answered that they are interested and that they can imagine a future participation in the project. 1.3% gave the information that they are interested but do not have a teacher who can supervise the project.

Schools who reported no interest in the project were asked for the main reason for this lack of interest. The schools who answered this question give reasons such as already having implemented a concept for robotics in education which would make the project redundant to them. Others state that the participation in the project would take too much effort.

Robots in the Classroom. Concerning the use of the robots in the classroom beyond the competition, we expected that most of the schools kept using the robots in their classes. A majority of 83.7% of the written survey participants who answered the question state that they do use the robots in the classroom beyond the competition.

Of the 16.3% that do not use the robots in classes 57.1% still use them in different educational contexts such as workgroups and projects. Data shows that the robots are used throughout all grades the different school types offer. In Germany, the school type *Realschule plus* offers classes from 5th to 10th grade whereas the *Gymnasium* additionally offers senior classes from 11th to 13th grade.

Throughout all school types, robots are also frequently used for projects that spread across grades. The non-participating schools were also asked whether they use robots in their lessons. 56.6% of those who answered the question do not use robots in their regular classes.

Given the opportunity to write a free answer on the question why they do not use LEGO® robots in their classes, 17.6% named costs as their main reason. With 11.8% each, the second leading causes are not having enough robots, staff shortage, not knowing about the project or robots that can be used in education, and robots being irrelevant because their school is specialized and feels that do not require the use of robots in their education.

The participating schools were also asked in which subjects they use the robots. Almost all schools reported that they use the LEGO® robots in various subjects. It can be observed that 38.9% of the survey participants who answered the question use the robots in regular subjects as technology education lessons, an elective subject at the *Real-schule plus*, and computer science, a subject that many *Gymnasiums* offer, but also in projects.

A majority of 72.2% reported the use of robots in workgroups. 22.2% of the schools that answered the question use robots in physics lessons. Some schools even created individual subjects for robotics and one school informed us that they have established special classes with emphasis on robotics. Of the schools that are not part of the project and who answered the question 58.8% reported the use of robots in workgroups. 29.4% use the robots in technology education lessons and 47.1% use them in computer science classes. 17.6% of the schools that answered to the question use robots in their physics lessons.

Here again, some schools gave the information that they have special classes in which they use the robots. Striking is that 100% of the participating vocational schools report the use of robots in their vocational training whereas only a third of the non-participating vocational schools who answered to the question use robots in their vocational training. In general, and despite very few exceptions, both the participating and non-participating schools use MINDSTORMS® in multiple subjects.

Implementation of Concepts. Concerning the concept for the implementation of robotics in education, that every project school has handed in with their application for the project, schools were asked if they still apply the concept. 82.8% of the schools that answered the question with "yes". Of those, 66.7% have already modified the concept and 20.8% still apply it in its original form. 12.5% are new to the project and are working on the implementation of the concept. Schools that are not part of the project "Robotics" have been asked whether they do have a concept for the use of robotics in their education or not. 75% of the non-participants that answered to the question do not have a sustainable concept for the use of robotics in education.

Implications for Learning and Teaching. All Schools were also asked to assess the learning effects of robotics in education with regard to four competencies: professional competence, social competence, media competence, and self-competence. For the professional, social, and media competencies, we follow the definitions of Kopf et al. [10]. Concerning the media competence, we follow the definition of Baacke [11]. They rated these on a Likert scale that starts with "1" (very great) and ends with "6" (very small). Concerning the professional competence, most answers accumulate on "2" (great). With respect to the social competence, again, most schools rated the learning effects as being big. The learning effects with relation to media competence received most ratings slightly better than average and the effects on the self-competence were, again, mostly rated with "2".

Noticeable is the overall similarity in the rating behavior between schools that are part of the project and schools that are not but still use robots. Both groups, independently from each other, give good ratings but the data shows that non-participants of the project estimate the learning effects throughout all competencies higher as the project schools do.

All schools were also asked to name the competency that benefits most from the use of robots in the classroom. Independently from the participation in the ministry's project, schools answered that the professional competency benefits most from the use of robots in the classroom. Additionally, project schools were asked to name the competency that they think might benefit the most from the participation in the FLL. With 46.4% of the answers, most project schools think that the social competence benefits the most from partaking in the FLL.

On the question whether the use of robots has enriched the teacher's and students' lessons, 81.3% of the non-participating and 79.3% of the project schools gave ratings of "1" and "2". Noticeable are the 10.3% of the project schools who gave ratings of "5" and "6" whereas none of the non-participants rated the enrichment of the lessons worse than "3". So, similar to the assessment of the learning effects, schools that do not participate in the project seem to be more satisfied with the use of robots in their lessons.

Of the project schools, 46.7% rated the students' motivation with "1" and 40% with "2". The average rating the participating schools gave their students' motivation when learning with robots is 1.7. Of the non-participants even 50% gave a "1" rating to the question about the motivation and 46.7% reported the motivation of the students as being good ("2"). With an average of 1.7, again, the non-participant group rate the motivation of the students during the learning process with robots better than project schools do.

Effects on the Frequency of Use of Robots in the Classroom. 44.2% of the schools that are part of the ministry's project reported that they possessed LEGO® robots before entering the project. Among the non-participating schools who answered the question, only 35% are in possession of LEGO® robots. However, 15% answered that they have robots from other manufacturers.

Of the schools that are part of the project, 55.6% answered that they have used robots in their classes before entering the project. 42.4% of the schools who are not part of the project report the use of robots in their classes.

Concerning the use of robots after entering the project, 35.3% of the schools who answered the question report that they use the robots more frequently now. A majority of 58.8% of the project schools answered that they use the robots with an equal frequency as they have done before entering the project "Robotics". A minority of 5.9% gave the information that they now use the robots less frequently.

Satisfaction of Participants with the Project. The participating schools also were supposed to rate to what extent the project has met their expectations. Of the survey participants who answered the question 22.5% gave the rating "1", meaning that their expectations have been fully met. With 42.5%, the rating "2" is the most frequently given grading. 15% of the persons who answered the question gave a rating of "4" and worse. The average rating is 2.5 which means that most expectations of the participants have been met by the project.

Even though not all the expectations of the participating schools have been met, 48.7% of people questioned who rated how recommendable the ministry's project is, gave a "1" rating, meaning that they would strongly recommend the project. Ratings of "4", "5", and "6" were checked by a minority of 2.6% each. The average rating is 1.9 which means that most of the participating schools would recommend the project "Robotics" to other schools.

4.3   Discussion of Results

Effects on Competence Development. A close look at the assessment of the learning effect, concerning the four already mentioned competencies shows that the rating behavior of both the participating and non-participating group who use robots in their lessons is very comparable but the ratings, concerning the professional and social competencies, of the non-participants are better than those of the project schools. Since the only significant difference between both groups is the participation in the project, there is no reason why their behavior should be different from each other.

However, this is no explanation for the fact that schools who do not participate in the project "Robotics" almost consistently give better ratings concerning learning effects with regard to professional, social, media, and self-competence. A possible reason might be the existence of a concept for the implementation of robotics in education.

All schools who partake in the project had to develop one and hand it in with their application for the project and 82.8% are still applying the concept in their lessons. In contrast, 75% of the schools that do not participate in the project and gave information about the existence of a concept for the implementation of robotics their lessons reported that they have not conceptualized the use of robots in their lessons, yet.

Assuming that a sustainable concept for the implementation of robotics in education would contain the formulation of learning goals that the students have to achieve, the absence of such a concept could be the reason for teachers to overstate the learning effects.

Concerning the use of robots, our data confirms the conclusions of Khanlari [4] and Benitti and Spolaôr [3], showing that 42.9% of the participating schools who reported that they use robots in their lessons think that the use of robots in regular classes mostly fosters the development of the professional competence.

59.4% of schools that do not participate in the ministry's project and who use robots in regular classes also estimate that the professional competence mostly benefits from the use of robots in the classroom. Eguchi [2] broadened this view and reported that learning with robots provides also opportunities for students to obtain content knowledge and acquiring critical academic skills.

A reason for the dominance of this particular competence in the answers of both groups can be found in the subjects in which the robots are used. Regarding regular lessons, both the participating and the non-participating schools reported that robots are mostly used in technology education lessons, computer science, and physics. These are subjects where the professional competency benefits most from the use of robots.

Possible evidence for this can be found in the results of Sullivan and Heffernan [12] and when comparing the assessment of both groups on the enrichment of the lessons by the use of robots with the answer on the question on which competence benefitting the most from the use of robots in regular classes. With 35.7%, the biggest number of schools who answered that the professional competence benefits the most from robots being used in regular classes rated the enrichment of the lessons through the use of robots with "1" and "2".

Implementation of Concepts for Robotics in the Classroom. With regard to what has become of the project of the Ministry of Education, we collected data on the implementation and modification of the concept that has to be handed in by every school with their application for the project. As mentioned before, the concept is still in use at 82.8% of the project schools.

We also collected data on how many years the concept is in use and whether the concept has been modified or is still being used in its original version. When looking at the answers, we suspected that the longer a concept has been in use, the likelier it has been modified. Unfortunately, we are not able to do a correlation analysis to test the hypothesis because we do not have information on the year in which the respective schools modified their concepts. However, data presented in the Table 1 indicates that a great number of schools start early with the modification of the concept. Hence, we suspect that most of the schools who modify their concepts start very early after the implementation. But, in order to test this, first, the respective data would have to be collected.

Table 1. Modification of the concept for the implementation of robotics in education

Years of application	% of schools	Schools that modified the concept	Schools that use the original concept
8	15.0%	15.0%	0.0%
7	25.0%	20.0%	10.0%
6	0.0%	0.0%	0.0%
5	5.0%	5.0%	0.0%
4	15.0%	15.0%	0.0%
3	25.0%	15.0%	10.0%
2	10.0%	10.0%	0.0%

We also suspected that the modification of the concept correlates with school lessons being more enriched through the use of robots. The initial concepts might contain points that need to be optimized in order to meet the demands of successful lessons. When looking at the collected information on the part of schools that altered their concepts it is clearly visible that the percentage of schools that modified their concepts is increasing with the extent to which the lessens have been enriched through the use of robots. In order to verify the thesis, the correlation coefficient has to show a reverse relation between schools who modified their concepts and the enrichment of the lessons.

Additionally, the correlation between schools that still use the original concept and the rating has to be either weak or non-existent. The calculation of Pearson's correlation coefficient proved a strong inverse relationship between the modification of the concept and the enrichment of lessons through the use of robots. As can be seen in

Table 2, the correlation between the use of the original concept and the enrichment of lessons is, indeed, reverse but very weak. As a result, the hypothesis is verified.

Table 2. Correlation analysis for the relation between the modification of the concept for the implementation of robotics in education and the enrichment of school lessons

"Did the use of robots enrich your lessons?"	Schools that modified the concept	Schools that use the original concept
1 (Yes, absolutely)	36.4%	9.1%
2	27.3%	4.5%
3	9.1%	0.0%
4	0.0%	0.0%
5	0.0%	4.5%
6 (Not at all)	0.0%	4.5%
Correlation coefficient	**−0.92**	**−0.36**

Reasons for Repeated Participation in the Contest. We also want to answer the question about the reasons for schools that are part of the project to keep partaking in FLL despite the fact that the state only accounts one time for the entry fee.

Table 3. Correlation analysis for the relation between the repeated participation in the FLL and the rating of the expectation being met by the project

"The project met my expectations"	Schools participating again in the FLL	Schools participating not again in the FLL
1 (Yes, absolutely)	23.1%	0.0%
2	26.9%	15.4%
3	11.5%	7.7%
4	0.0%	3.8%
5	0.0%	3.8%
6 (Not at all)	3.8%	3.8%
Correlation coefficient	**−0.86**	**−0.19**

The hypothesis is that the more the expectations of new schools are met by the project, the more likely the schools will again participate in the FLL. In order to prove the hypothesis to be true, again, Pearson's correlation coefficient has to show an inverse relationship between the two variables.

Furthermore, the relationship between schools that will not again participate in the FLL and the project meeting the expectations needs to be either parallel of non-existent. The analysis revealed a strong inverse relationship between the expectations being met by the project and schools participating again in the FLL.

As seen in Table 3, even though the correlation coefficient for the second condition is negative, the correlation is very weak. This means that the hypothesis can be confirmed. Schools whose expectations have been met by the project tend to repeatedly participate in the FLL.

Effects on the Use of Robots in the Classroom. Another question is the impact of the participation in the project and the frequency in which robots are used in comparison to the time before the school entered the project.

Schools were asked the question whether they use robots more frequently, less frequently or if the use of robots remains unchanged compared to the time before the project. 35.3% of the schools who answered the question reported a more frequent use after entering the project.

Surprisingly, the majority of 58.8% gave the information that their use of robots remains unchanged and a minority of 5.9% uses the robots less frequently since their participation in the project. Despite the fact that the participation in the ministry's project did not change the frequency in use for most schools, still more schools report a more frequent use of robots that schools did report that they use them less frequently.

However, the collected data does not contain any information about the actual frequency of use before and after the participation in the project. It just shows a trend that the project seems to have a positive impact on the frequency of use of the robots. This is due to the fact that more schools reported a more frequent use than schools who reported to use the robots now less frequent. The school which reported that they now use the robots rarer, stated that their concept for the implementation did not prove to be successful when put into practice.

Potential of the State's Program. Ultimately, we were interested in the potential for the acquisition of new project schools among the non-participants who do not know about the project "Robotics".

To achieve this, we subtracted the schools that do not know about the project in order to be able to calculate the proportions of schools who are interested and schools who are not. Assuming that all schools who do not know the project ticked the box with the corresponding answer, we can presume that the remaining schools are either interested or uninterested in the participation in the project.

The former group constitutes 73.7%, the latter 26.3% of the remaining schools. This ration can now be applied to the absolute number of 293 schools who do not know about the project. As a result, the potential for the acquisition of new project schools, 73.7% of the 293 schools, meaning, 216 schools could be interested in the project and potentially be won for taking part in it. However, this calculation is not exact because of an uncertain number of schools that consider themselves as schools which do not have anything to do with robotics like, for instance, schools that provide training in nursing care for the elderly.

Since we do not know the exact number but are sure that, in the age of digital-ization, even students of those schools could benefit from the use of robots, we did not try to exclude them but rather assumed that they would be among the schools who checked the box for those who are not interested in the project.

5   Conclusions and Implications

All in all, the survey shows that most schools benefit from participating in the project "Robotics" by the Ministry of Education. Most of the project schools tend to participate in the FLL beyond the project. In average, the use of robots in regular classes increases after entering the ministry's project and shows a more successful way to achieve a higher level of participation in educational robotics program when compared to the methods of getting schools involved in robotics projects in Estonia reported by Altin et al. [13]. Even though the robots are mostly used in workgroups, they also enrich regular lessons in subjects like technology education lessons, computer science, physics, and more. Not to forget the individual classes for the use of robotics that some schools created.

Even though schools rate the preparation time for robot lessons only slightly better than average, they report that the use of robots enriched their classes to a high extend. The collected data also reveals that the project meets the expectations of most of the participating schools. Additionally, a great majority of the participants regards the project as highly recommendable.

This may be due to the implementation and modification of sustainable concepts for the use of robotics in education because the study revealed that the schools who modify their concepts are also the ones that give the highest ratings when asked to which extend the use of robots has enriched their lessons.

The data also provides information on the obstacles that keep schools from using robots or entering the ministry's program. Schools name costs, a shortage of hardware and personnel, as well as the effort that is associated as reasons that keep them from implementing robots in education.

Also, there are schools that cannot relate to technology education. The biggest obstacle that keeps schools from participating in the project "Robotics" is that half of the non-participating schools are not aware of the project's existence.

The calculation of the project's potential among these group indicated approximately 200 schools which could be engaged in the use of robotics or who could potentially better themselves in that field. Thus, our findings confirm Altin et al. [13] in their recommendations that more extensive marketing on the project would increase the implementation of robotics in education on a larger scale.

Like the conclusions of Benitti and Spolaôr et al. [3], the study indicates that schools benefit to a great extend from having a sustainable concept for the implementation of robotics in education. Additionally, schools who modify the concepts give excellent ratings on the enrichment of their lessons through the use of robots. This would be a link to further studies that could be conducted. It would be of great interest to collect and analyze information on the existing concepts and in which ways schools did alter them in order to meet the demands of their teachings to, eventually, find out what matters in terms of great concepts for the implementation of robotics in education.

References

1. Ministerium für Bildung. Grundkarte: "Projekt 'Robotics'"
2. Eguchi, A.: Bringing Robotics in Classrooms. In: Khine, M.S. (ed.) Robotics in STEM Education: Redesigning the Learning Experience, pp. 3–31. Springer, Heidelberg (2017)
3. Benitti, F.B.V., Spolaôr, N.: How have robots supported STEM teaching? In: Khine, M.S. (ed.) Robotics in STEM Education: Redesigning the Learning Experience, pp. 3–31, 124. Springer, Heidelberg (2017)
4. Khanlari, A.: Effects of robotics on 21st century skills. Eur. Sci. J. 9(27), 26–36 (2013)
5. Alimisis, D.: Educational robotics: open questions and new challenges. Themes Sci. Technol. Educ. 6(1), 63–71 (2013)
6. Nugent, G., et al.: Robotics camps, clubs, and competitions: results from a U.S. robotics project. In: Alimisis, D., Granosik, G. (eds.) Proceedings from the 4th International Workshop Teaching Robotics: Teaching with Robotics & 5th International Conference Robotics in Education, Padova, pp. 11–18 (2014)
7. Karp, T., Maloney, P.: Exciting young students in grades K-8 about STEM through an afterschool robotics challenge. Am. J. Eng. Educ. 4(1), 39–54 (2013)
8. Hands on Technology e.V. "What is FLL" Website – FIRST® LEGO® League. https://www.first-lego-league.org/en/general/waht-is-fll.html
9. Hands on Technology e.V. "Outline of FLL" Website – FIRST® LEGO® League. https://www.first-lego-league.org/en/general/waht-is-fll.html#5
10. Kopf, M., et al.: Kompetenzen in Lehrveranstaltungen und Prüfungen: Handreichungen für Lehrende. ZQ (2010)
11. Baacke, D.: Handbuch Medien: Medienkompetenz: Modelle und Projekte. Bundeszentrale für politische Bildung (1999)
12. Sullivan, F.R., Heffernan, J.: Robotic construction kits as computational manipulatives for learning in the STEM disciplines. J. Res. Technol. Educ. 48(2), 105–128 (2016). https://doi.org/10.1080/15391523.2016.1146563
13. Altin, H., Pedaste, M., Aabloo, A.: Robotics in education: methods of getting schools involved in robotics project in Estonia. In: Proceedings of SIMPAR 2010 Workshops. International Conference on Simulation, Modeling and Programming for Autonomous Robots, Darmstadt, 15–16 November 2010, pp. 421–428 (2010)

Improving Students' Concepts About Newtonian Mechanics Using Mobile Robots

Paola Ferrarelli[1]([⊠]), Wilson Villa[1], Margherita Attolini[2], Donatella Cesareni[2], Federica Micale[2], Nadia Sansone[2], Luis Claudio Pantaleone[3], and Luca Iocchi[1]

[1] Department of Computer, Control, and Management Engineering,
Sapienza University of Rome, Rome, Italy
ferrarelli@diag.uniroma1.it
[2] Department of Social and Developmental Psychology,
Sapienza University of Rome, Rome, Italy
[3] Department of Chemistry, Sapienza University of Rome, Rome, Italy

Abstract. In this paper we proposed an educational robotics project using MARRtino mobile robot to students of Italian high schools, aged from 15 to 19 years old. The MARRtino Educational Robot for high school project (MHS) lasted 70 h during the 2017/2018 school year. Students teams built a mobile robot and programmed it to participate to RomeCup 2018 competition and to make experiments in their school Physics Laboratory. We measured if the educational robotics activity improved students Physics concepts learning, Technology perceptions and attitudes, and perception of collaborative work attitudes. Moreover, the by-product of the educational activity was to broaden participation of young students in the technology field and to enrich the high school Physics Laboratories with a new technological and educational tool, a mobile robot.

1 Introduction

A survey on the use of robots in educational field was made by Benitti [1], who reviewed the international literature on Educational Robotic (ER) published over 10 years listing used robots, students age and obtained results. From the literature analysis, it emerged that many of ER activities were extra-curricular, during summer camps or workshops, few studies showed quantitative data, using statistical methods, or analyzed the use of the robots to teach subjects different from computer science or mechatronics, finally most of the research was focused on kids (4–10 years). Mubin [10] gave an overview on robot kits, robot roles and robot usage domains. About robot kits, a recent study was conducted by Garcia [16] by listing characteristics of robot kits and apps, currently available on the market for teachers of 4–14 age students, and costs, that for Mondada [9] was one of the obstacles to the massive use of robots in the schools. About robot role, Shin [18] analyzed the interviews to 85 students highlighting that young students

© Springer Nature Switzerland AG 2019
W. Lepuschitz et al. (Eds.): RiE 2018, AISC 829, pp. 113–124, 2019.
https://doi.org/10.1007/978-3-319-97085-1_12

preferred a robot that acted as a peer during the learning process. About robot usage, the use of robots in a class was not the only favorable condition to the learning process, but also: the presence of the teacher, the spaces suitable to do the activities with the robot, the availability of one kit for each team of 2–3 students, short theory lessons and tutorials to link theory and practice, realistic but affordable tasks linked with curricular subjects, teachers at ease with the robot, etc. In conclusion, ER researchers and teachers need to choose the more suitable robot kit for their students, and carefully design where and how to use it and with which role.

We decided to investigate the possibility to use robotics to teach Physics Newtonian concepts to high school students, aged 15–19 years old. The researchers in physics education [2, 4, 7], showed that traditional passive-student physics courses generate little conceptual understanding of Newtonian mechanics, even if they were delivered by the most talented instructors. Using massive pre- and post-course testing of students, Hestenes and Halloun [6] concluded that students misconceptions about motion were hard to overcome because they were grounded in their personal experience. The authors presented a taxonomy of common sense as a guide for instructional design [5], to be used by physics teachers. Hestenes et al. [8] introduced an instrument (the Force Concept Inventory) to help teachers probe and assess the common senses beliefs of their students. By the way, knowing the student misconceptions was not sufficient by itself to improve the effectiveness of the instruction. To induce the conceptual change, an ad-hoc instructional method must be designed and developed.

The purpose of this paper was to illustrate an instructional method, based on the use of a mobile robot, to improve high school students' concepts about Newtonian mechanics. The effectiveness of our method was tested using the Force Concept Inventory (FCI). Moreover, we studied the success of the method in terms of students' technology perceptions and attitudes and their perception of collaborative work attitudes. The research methodology was based on pre- and post-course testing and the results of these tests are illustrated in this paper.

2   Research Methodology

Our research wanted to explore three different domains: the learning of Physics concepts described in [8], the technology perceptions and attitudes [19], the perception of collaborative work attitudes [14]. Results in the three domains were measured using surveys that were administered at the beginning (pre-test) and at the end (post-test) of the project.

The participants were students coming from different high schools around Rome city, aged from 15 to 19 years, attending the last 3 years of high school (III, IV and V). A group of 113 students, including 24 females and 89 males, participated to the MHS project, started on November 2017 and described below (Robotics group). In order to have a feedback on the MHS results, we took into consideration a group of 96 students (including 35 females and 61 males), participating to project "Lab2go" [12]. Lab2go project, started on November

2017, had the same organization and users typology of the MHS project, except that the activities were related to physics not robotics (Physics group). Pre- and post-test data, related to the two groups of students, were analyzed to observe if the MHS project changed some of the three domains under investigation.

2.1   MHS Project

The methodology for the MHS project was a long-term hands-on course for small groups of high school students (4/5). The course, started on November 2017, lasted 70 h during the 2017/2018 school year. There were two goals: RomeCup 2018 competition (NonniBot contest)[1] and Physics Laboratory experiments[2]. The MHS activities were done at the University (40 h), at school (20 h) and at home (10 h). At the University, students spent 4 h once a week, and they were supported by university professors and undergraduate students (expert tutors). Here the students built the robot (Fig. 1), installed the software required to control the robot, learned basic knowledge of Python programming and how to run a simulator. Then, the activities were continued in the schools, were the students developed autonomously their experiment, receiving remote assistance by expert tutors, and wrote the documentation about it on the project website. Finally, there were activities done at home, like writing code or web pages to document the project. Students recruitment was done by high school teachers that adhered to the MHS project published by Sapienza University. High school students decided to apply to the MHS project, on a voluntary base. No background knowledge was required for the students, except their availability to participate to weekly meetings (morning or afternoon) at the University or at their school. The role of the school teachers was organizational: they organized the meetings at school and communicated meetings organized at the University to their students. If they wanted, they were welcome to follow the project activities.

Fig. 1. Students assemble the components to build MARRtino Educational Robot.

---

[1] http://www.romecup.org/competizioni.

[2] http://www.roma1.infn.it/LAB2GO/informatica-robotica/index.html.

Table 1 summarized the characteristic of the MHS project in terms of number of participants, robot role and cost, domain and location of the learning activities, that were fundamental parameters to describe the experiment, as underlined by [10]. All the teams of students built and calibrated their robots, they learned how to program it and they developed Python programs to implement the project they will demonstrate either during the RomeCup2018 Nonnibot competition or in the Physics laboratory.

Table 1. Parameters of MHS project.

Number of MHS students	113
Where	University, school, home
Robot role	Tool to engage students to learn
Skills	Physics, Technology, Collaboration
Teacher role	Organizational, learner
Univ. Professor role	Expert tutor
Background knowledge	Not required (both for students and teachers)
Robot cost	Low cost product, affordable for public schools

2.2   MARRtino Educational Robot

MARRtino[3] is a simple differential drive robot and its main components are: a simple box chassis made of plexiglass, an Arduino Mega 2560 used together the Arduino Motor Shield in order to control the two motors and to read data coming from the encoders, a 12 V battery, a small LED display to show the battery voltage, a switch to turn on/off the robot, a charger and a PC or Raspberry Pi to program it. The robot can speak, it can receive commands through voice from a phone, it is equipped with a camera that can be used to detect predefined tags. Students assembled the robot and tested that it could move accordingly to simple commands sent (i.e. move left wheel). MARRtino is a ROS-based robot, thus can exploit all the available open-source components from ROS community[4]. However ROS was hidden to the students thanks to the use of high level user-friendly interfaces. The students indeed learned to write Python programs with simple high level commands (e.g., `forward()`, `left()`, ...) and they could exercise with a simulated robot based on Stage [20]. The programming experience was made more user-friendly using Blockly, an open source library that made it easy to add block based visual programming to an app [3,15] (an example is shown in Fig. 2). Once the developed program is verified on the simulator, it could be sent to the MARRtino robot using the web interface, by just inserting the IP of the robot, connecting to it, and running the program.

---

[3] https://tinyurl.com/marrtino.
[4] www.ros.org.

We proposed to the students to program physics experiments like: program the robot to move 1 m forward then measure how much the robot has moved; move the robot at constant velocity and observe if a ball above it will move; accelerate the robot and observe in which direction the ball will move and measure the displacement; provide the robot with a pen and move it forward and backward, like a pendulum, on a sliding sheet of paper, to draw a wave and measure the frequency of the motion; verify the Doppler effect using 2 robots: one emitting a sound and moving, the other standing with an oscilloscope and a microphone above it; collect robot velocity at constant time's intervals and draw a graph of the motion; measure the minimum speed at which the robot goes straight on a circular path. All these experiments exercise the 6 conceptual dimensions verified by the FCI questionnaire (see later).

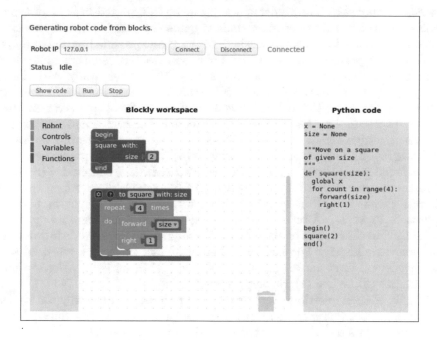

Fig. 2. Programming experience using Blockly.

2.3   Surveys

Results in the three domains were measured using surveys that were administered at the beginning (pre-test) and at the end (post-test) of the project. Pre and post data, related to the Robotics and Physics groups of students, were analyzed to observe if the MHS project changed some of the three domains under investigation.

Knowledge about Physics concepts was assessed using the Force Concept Inventory v95 (FCI), a questionnaire with 30 multiple choice questions [8]. FCI survey was chosen because it assesses students' understanding of the most basic concepts in Newtonian physics (forces, kinematics), concepts that can be easily explored using the mobile robot like MARRtino. It is characterized by having exactly one correct answer for each question; it assesses students' understanding of the most basic concepts in Newtonian physics, using everyday language and common-sense distractors; it is appropriate for Intro college and High school students. FCI was analyzed using the methodology described in previous works [8]. This test includes clusters of questions by topic, that help to have a better sense of which conceptual dimension the students need more help with. The topics are: Newton first, second and third laws, kinematics, superposition and kind of forces. FCI administration and scoring was done following the PhysPort Expert Recommendation on Best Practices for Administering Concept Inventories[5]. We used the PhysPort Assessment Data Explorer for analysis and visualization of students' responses[6]. The students' score can be interpreted in light of typical results [21] to get a sense of what kinds of gains are possible in different kinds of courses. We used the Italian version of the survey, provided by the authors.

Technology perceptions and attitudes were assessed with a questionnaire composed by 11 items, readapting the one used by [19] in their study. After Factorial analysis, the items were grouped in 4 scales: Attitude towards Technology, Attitude towards the University, Perception of the Technology Sector and Self-Efficacy in Technology Sector respect to Gender. For each item there was an agreement/disagreement scale with five-level Likert modes: Strongly disagree (1), Disagree (2), Neither agree nor disagree (3), Agree (4), Strongly agree (5).

Students' perception of collaborative work attitudes was collected with a questionnaire composed by 20 items, re-adapted from the CKP questionnaire (Collaborative Knowledge Practices [11]), that is inspired by the Trialogical Learning Approach [13,14,17]. After Factorial analysis, the 20 items were grouped in 5 scales: Collaboration; Use of Technology; Willingness to improve their own work; Autonomy; Reflexivity. For each item there is an agreement/disagreement scale with five-level Likert modes: Strongly disagree (1), Disagree (2), Neither agree nor disagree (3), Agree (4), Strongly agree(5).

Finally, 2 open questions were used in the pre-test to know why the student chose the project (curiosity, passion) and how (was a choice or a duty); 2 open questions were used in the pre-test to know the student's expectations (about the project contents and learning) and in the post-test to know if the expectations have been met.

3   Data Analysis Results

For both groups under test (Robotic and Physics), we analyzed: the physics questionnaires (pre-test only, since currently students are deploying their physics experiments and the post-test will be administered when the projects finish); the technology and collaborative work attitudes surveys (both pre- and post-test),

---

[5] www.physport.org/expert/AdministeringConceptInventories/.

[6] www.physport.org/explore/FCI.

using suitable statistical methods; the post-test open questions of the collaborative work attitudes survey, that measure students expectations and their perception of what they have learned (only qualitatively). Table 2 sums up the available
data analysis.

Table 2. Surveys: what we have analyzed. In the table, "o.q." means "only qualitatively".

Group	Robotics		Physics	
	Pre	Post	Pre	Post
Physics concept of motion	Yes	No	Yes	No
Technology attitudes	Yes	Yes	Yes	Yes
Collaborative work attitudes	Yes	Yes	Yes	Yes
Collaborative work attitudes open questions	o.q.	o.q	o.q.	o.q

3.1   Physics Survey Results

We administered the FCI v95 to 209 students participating to the Robotics and
to the Physics projects.

Since at the moment we have only pre-test scores, we can deduce students'
initial ideas about force and motion.

169 students answered the pre-test (82 Robotics and 87 Physics). The overall
pre-test mean score of correct answers, 39% (std.dev. $= 19$), was quite low, but in
line with similar results obtained by [21] on a larger sample of students pre-tests.

The analysis of correct answers, clustered by topics, shows that the most
impacted areas (less than 40%) are: Newton second and third laws, kind of
forces and superposition (see Table 3). Newton's first law and kinematics are
better understood (44% both). Question 13, 15 and 30 were the hardest item of
the test. Item 13 regards gravitation and it is related to conceptual dimension
6. It lends itself to analysis by force diagram for identification of forces. You
can see the pie chart in Fig. 3 describing how answer choices were distributed.
The black sector represents the correct answer (14%), while the white sectors
represents the distribution of the wrong answers. We can see that not only are
many students missing the question, but 54% of them are consistently selecting
the same wrong answer C. The selection of this option reveals the presence of
the misconception called *impetus* (its dissipation in this case). The impetus is
conceived to be an inanimate motive power that keeps things moving, contra-
dicting Newton first law. Item 15 investigates conceptual dimension 3, related to
action/reaction pairs. As shown in Fig. 3, 64% of students answered C, revealing
the misconception *most active agent produces greatest force*. Question 6 and 7,
related to conceptual dimension 1 (Newton first law with no force), were the
easiest items, both with 71% of correct answers.

The analysis performed so far shows that the results are in line with previous
works where students do not use robots in physics laboratory [21].

Table 3. FCI pre-test: mean correct answers percentage, clustered by topics.

Conceptual dimensions	Mean correct (%)
1. Newton first law	44
2. Newton second law	38
3. Newton third law	35
4. Kinematics	44
5. Superposition	37
6. Kinds of forces	37

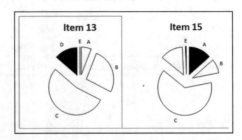

Fig. 3. Item 13 and 15 pie charts. The black sector represents the percent of correct answers.

We can drill down data to see if there were differences between Robotics and Physics group of students.

Robotics Group: 82 students completed the pre-test. The overall pre-test mean score of correct answers, was 33% (std.dev. = 19). The analysis of correct answers, clustered by topics, showed a distribution of data spanning from 26% to 42%. The most impacted areas (less than 30%) were: Newton third laws, kind of forces and superposition. Newton's first law and kinematics were better understood (37% and 42%). Questions 5 and 13 were the hardest items on the test, revealing misconceptions on dimension 6.

Item 4 had the greater percentage of wrong answer (71%). It investigates conceptual dimension 3, related to action/reaction pairs and the most of students had the misconception *greater mass implies greater force*. Items 6 and 9, related to conceptual dimension 1 (Newton first law with no force) and 2 (impulsive force and vector addiction of velocities), were the easiest items, both with 65% of correct answers.

Physics Group: 87 students completed the pre-test. The overall pre-test mean score of correct answers was 45% (std.dev. = 21). The analysis of correct answers, clustered by topics, showed a quite homogeneous distribution of data spanning from 43% to 50%. Item 30 was the hardest item of the test, related to conceptual dimension 6: 59% answered wrong, revealing the misconception *impetus supplied by hit*. Item 15 had the greater percentage of wrong answer (77%). It investigates conceptual dimension 3, related to action/reaction pairs, revealing the misconception *most active agent produces greatest force*. Items 7, related to conceptual dimension 1 (Newton first law with no force), was the easiest item, with 79% of correct answers.

3.2 Technology Perceptions and Collaborative Work Attitudes Surveys Results

In this experiment we analyzed the data regarding changes in the technology perceptions and collaborative work attitudes, reporting for each identified scale the mean, the standard deviation (Std.Dev.), the one-way ANOVA F (F) and the significance level (sign.) values. All the results were summarized in Tables 4 and 5, where bold style identifies significant data.

Technology Perceptions. The pre- and post-test analysis showed that scores were generally high (4 = Agree, 5 = Strong Agree). This was probably related to the fact that 89% of students chose the project between others proposed by high school (pre-test question 12) and 65% of them selected the project because interested in it while the others out of curiosity (pre-test question 13).

There were no differences between the Robotics and Physics groups. Considering differences between pre and post test, we could see that there were significant higher mean values in the "Attitude towards the University" scale.

At the end of the activity students were more likely to enroll at the University.

We observed also significant differences in the "Self-Efficacy in Technology Sector respect to Gender" scale, where a lower value in the post-test means that the perceived gap between the "use of technology" and the "gender" reduced, supporting the idea that technology and scientific subjects (STEM) were both for female and male.

We observed lower post-test mean scores regards "Attitude towards Technology" and "Perception of the Technology Sector", but were not significant. The analysis of the open questions, in future papers, could help us to better understand this phenomenon.

Collaborative Work Attitudes. The data analysis for collaborative work attitudes, reported in Table 5, showed high average scores assigned to all the scales. The students therefore showed, in general, a high perception of their own Collaborative Knowledge Work skills. As already mentioned, this fact was influenced by the students' particular interest to participate to these projects. There were no significant differences between Robotics and Physics groups. The post-test mean values were higher than the pre-test ones, for all the scales but they were not statistically significant, probably because they were very high in the pre-test, already. The "Autonomy" scale value was near to be significant.

Collaborative Work Attitudes Open Questions. From a first qualitative analysis of the collaborative work attitudes post-test open questions seems that the expectations of the students were fully and partially realized in 60% and 26% of cases. 36% perceived to have learned how to work in group (while in the pre-test, only 10% perceived it as an expectation and 6% perceived they would

Table 4. Technology perceptions and attitudes pre- and post-test data. The bold style identifies significant data.

Time	Pre		Post		Differences
	Mean	Std. dev.	Mean	Std. dev.	F (sign.)
Attitude towards technology	4.3149	.52070	4.2155	.61432	2.985 (.085)
Perception of the technology sector	3.7372	.62127	3.6667	.81271	.935 (.334)
Attitude towards the university	**4.4327**	.65380	**4.5746**	.69818	**4.276 (.039)**
Self-efficacy in technology sector respect to gender	**1.7067**	.93538	**1.5166**	.80605	**4.544 (.034)**

Table 5. Perception of collaborative work attitudes pre- and post-test data.

Time	Pre		Post		Differences
	Mean	Std. dev.	Mean	Std. dev.	F (sign.)
Collaboration	3.9962	.64760	4.0867	.64295	1.906 (.168)
Use of technology	3.9724	.70507	4.0622	.65134	1.685 (.195)
Willingness to improve their own work	3.9808	.67000	4.0295	.60372	.560 (.455)
Autonomy	4.0629	.59789	4.1727	.57454	3.383 (.067)
Reflexivity	4.0335	.60683	4.1046	.64133	1.261 (.262)

have learned team work). Moreover 35% perceived to have acquired or improved technical skill (while in the pre-test, nobody expressed it as an expectation and only 8% perceived they would have learned technical skill). In a future paper we will analyze them deeply[7].

3.3   Discussion

The analysis of the pre- and post-tests showed its utility in evaluating and assessing our instructional methodology. The results obtained so far provide a clear picture of the final point of the users involved in this experimentation, showing several significant perceptions and attitudes. We filled some gaps evidenced in Benitti [1] research, because we assessed our methodology using quantitative data, treated in a statistical way. Moreover, we addressed a curricular subject (Physics), involving 15–19 years old students during lesson time at school (no extracurricular activities). We also implemented best-practices evidenced in Mubin [10] research: one robot kit for small team of students (team work accelerator); realistic but affordable tasks (move the robot forward/backward); short

---

[7] Web site of the MHS project with additional information, material and results: https://sites.google.com/a/dis.uniroma1.it/asl-robot/MHS_Rie2018.

theory lessons; suitable spaces for robot experiments (school and University laboratories); the presence of the school teacher (organizational role). Finally, to asses the improvement in physics knowledge, we used the FCI survey, that was used on large samples by [21] and we found that students initial ideas about force and motion, were consistent with the one discovered in previous works.

4   Conclusions

In this paper we presented an instructional methodology, using MARRtino mobile robot. We applied it to a group of 113 students coming from 14 different high schools around Rome city that chose to participate to MHS project (Robotics group). We used pre- and post-test methodology to evaluate and assess our instructional method.

The questionnaires explored three domains: students changing in their physics concepts of motion, technology perceptions and attitudes, and the perception of collaborative work attitudes.

In order to have a feedback on the MHS results, we administered the same questionnaires to a group of 96 students participating to the Physics project (Physics group), that had the same organization and users typology of the MHS project. Data, related to the two groups of students, were analyzed to observe if the MHS project changed some of the three domains under investigation.

From the analysis of the pre- and post-test questionnaires of the perceptions of technology, we found significant improvements in "Attitude towards the University" and in the "Self-Efficacy in Technology Sector respect to Gender" scales, with no difference between the Robotics and Physics group under investigation.

From the analysis of the pre- and post-test questionnaires of the collaborative work attitudes, we found improvement in all the scales, with no difference between the Robotics and Physics group under investigation and the "Autonomy" scale value near to be significant.

From the analysis of the pre-test questionnaires of physics concept of motion, we discovered the initial nature and extent of physics misconceptions in the Newtonian domain of the two groups. The analysis performed showed that the results were in line with previous works where students did not use robots in the physics laboratory.

Future works will consist in collecting post-test data on the physics concept of motion, compare them with the pre-test, and eventually analyze the collaborative work attitudes open questions.

Acknowledgements. We would like to express our appreciation to all the students who adhered to MHS project and the teachers that supported them.

References

1. Benitti, F.B.V.: Exploring the educational potential of robotics in schools: a systematic review. Comput. Educ. **58**(3), 978–988 (2012)

2. David, H.: A role for physicists in stem education reform (2015)
3. Fraser, N.: Ten things we've learned from blockly. In: 2015 IEEE Blocks and Beyond Workshop (Blocks and Beyond), pp. 49–50, October 2015
4. Hake, R.R.: Interactive-engagement versus traditional methods: a six-thousand-student survey of mechanics test data for introductory physics courses. Am. J. Phys. **66**(1), 64–74 (1998)
5. Halloun, I.A., Hestenes, D.: Common sense concepts about motion. Am. J. Phys. **53**(11), 1056–1065 (1985)
6. Halloun, I.A., Hestenes, D.: The initial knowledge state of college physics students. Am. J. Phys. **53**(11), 1043–1055 (1985)
7. Hestenes, D.: Wherefore a science of teaching? Phys. Teach. **17**(4), 235–242 (1979)
8. Hestenes, D., Wells, M., Swackhamer, G.: Force concept inventory. Phys. Teach. **30**(3), 141–158 (1992)
9. Mondada, F., et al.: Bringing robotics to formal education: the thymio open-source hardware robot. IEEE Robot. Autom. Mag. **24**(1), 77–85 (2017)
10. Mubin, O., Stevens, C.J., Shahid, S., Al Mahmud, A., Dong, J.-J.: A review of the applicability of robots in education. J. Tech. Educ. Learn. (2013)
11. Muukkonen, H., Ilomäki, L., Lakkala, M., Toom, A.: Adaptation of the collaborative knowledge practices questionnaire to upper secondary education. In: EDULEARN 2017 Proceedings, a Paper Presented at the 17th Biennial Conference for Research on Learning and Instruction (EARLI), 29 August–2 September (2017)
12. Organtini, G., et al.: Promoting the physics laboratory with lab2go. In: EDULEARN 2017 Proceedings of 9th International Conference on Education and New Learning Technologies, 3–5 July 2017, pp. 5264–5268. IATED (2017)
13. Paavola, S., Hakkarainen, K., Sintonen, M.: Abduction with dialogical and trialogical means. Logic J. IGPL **14**(2), 137–150 (2006)
14. Paavola, S., Lipponen, L., Hakkarainen, K.: Models of innovative knowledge communities and three metaphors of learning. Rev. Educ. Res. **74**(4), 557–576 (2004)
15. Pasternak, E. Fenichel, R., Marshall, A.N.: Tips for creating a block language with blockly. In: 2017 IEEE Blocks and Beyond Workshop (B&B), pp. 21–24, October 2017
16. Rees, A.M., García-Peñalvo, F.J., Toivonen, T., Hughes, J., Jormanainen, I., Vermeersh, J.: A survey of resources for introducing coding into schools (2016)
17. Sami, P., Kai, H.: From meaning making to joint construction of knowledge practices and artefacts: a trialogical approach to CSCL. In: Proceedings of the 9th International Conference on Computer Supported Collaborative Learning, vol. 1, pp. 83–92. International Society of the Learning Sciences (2009)
18. Shin, N., Kim, S.: Learning about, from, and with robots: students' perspectives. In: The 16th IEEE International Symposium on Robot and Human interactive Communication, RO-MAN 2007, pp. 1040–1045. IEEE (2007)
19. Sowells, E., Waller, L., Ofori-Boadu, A., Bullock, G.: Using technology summer camp to stimulate the interest of female high school students in technology careers. In: 2016 Portland International Conference on Management of Engineering and Technology (PICMET), pp. 1688–1696. IEEE (2016)
20. Vaughan, R.: Massively multi-robot simulation in stage. Swarm Intell. **2**(2), 189–208 (2008)
21. Von Korff, J., et al.: Secondary analysis of teaching methods in introductory physics: a 50 k-student study. Am. J. Phys. **84**(12), 969–974 (2016)

Multigenerational Collaboration to Create a Community of Practice Through Robot Application Development

Nahoko Kusaka[1]([✉]), Nobuyuki Ueda[1], and Koichi Kondo[2]

[1] Doshisha Women's College of Liberal Arts, Kodo, Kyotanabe, Kyoto, Japan
nkusaka@dwc.doshisha.ac.jp
[2] Fubright Communications, Inc., 1-20-1, Ginza, Chuo-ku, Tokyo, Japan

Abstract. The purpose of this study was to examine the effects of college classroom-based collaborative learning by exploring robot application on the interactions between the elderly, college students and robots, and social competence among participants. The trial included eight elderly people and 12 female college students in one class. Participants completed questionnaires before and after participation in the course. To analyze the process of co-creation of robot application, communication between the elderly and college students was observed.

In the first trial of the collaborative learning program using a robot with the elderly and college students, during the discussion of the class, an increase of participants' synchronicity was observed and recognized. The participants' attitude toward the future changed to a more positive direction after participation. These results indicate collaborative learning using a robot appears to improve inclusive interaction among multi-generations and the robot.

Keywords: Robot application · Project-based learning
Multi-generational interaction · Collaborative learning

1 Background

Social isolation in old age is a serious problem in many developing nations where life expectancy and the number and proportion of older people are rapidly increasing. Based on clinical and empirical data, social isolation has negative consequences on older people's physical and psychological health. In Japan, there is a traditional stereotype that elderly are those who are supported and younger people are the supporters, and this social condition impedes intergenerational exchanges.

In order to solve the problem of isolation, developing the effective intervention to destroy these fixed human relationships and build a new relationship that transcends all generations is needed. We have been engaging in a community of practice [2] to encourage autonomous social interactions with older adults [3, 4], however effective learning opportunities for older people to cope with aging and the rapid changes in society (e.g., changes in social interactions, their ability to engage in society as their capacity declines, rapid innovation in society) are scarce. All generations, including young people, are required to improve their social skills to coexist in harmony with

© Springer Nature Switzerland AG 2019
W. Lepuschitz et al. (Eds.): RiE 2018, AISC 829, pp. 125–136, 2019.
https://doi.org/10.1007/978-3-319-97085-1_13

other generations and new technology-like robots to stay involved in the community. The OECD's Definition and Selection of Competencies (DeSeCo) indicated that three key competencies: interacting in heterogeneous groups, acting autonomously and using tools interactively are needed for a successful life and for a well-functioning society. Improving these competencies is necessary to form an inclusive society, including multi-generations and robots.

2 Purpose

In the present study, humanoid communication robots were placed into the groups in order to breakdown the existing relationship between older people and young people and to establish a new relationship. The college class-based collaborative learning was established to create autonomous social interaction in the community.

Previous research of robot psychology tended to focus on dyad relations between a person and robot. However, more than two people or three agents consist the community and robots will affect the community in modern society. It is necessary to explore the influences of robots on the relationship of multi-generational interactions, when robots are put into a human community, to find an effective way of using the humanoid robot to solve the problems of isolation. Heider's balance theory [1] can be applied to explore the communication in the community including the robots. Heider's balance theory which shows the balance on the relationship between three things: the perceiver, another person, and an object can be used as a hypothesis in considering what kind of interaction occurs in the relationship between elderly people - college students - a robot. Changing feelings toward other people or robots in a tripartite dialogue can be captured by intentionally controlling how the robot responds and reacts, and subsequently the emotional changes among the three can be observed.

This study examined the effectiveness of the class as it relates to learning outcomes in older people and female students attending a liberal arts university. The middle-aged, elderly people and young women who are not tech savvy and lack sufficient computer or digital technology skills are becoming more and more anxious in this growing world of robotics. Although the elderly and the students may initially feel anxious about being involved in a project using robots and digital devices, the project gives them an opportunity to gradually move from being passive users of high tech devices towards being innovative and creative developers. While the two groups may feel uneasy about using technology, their anxiety encourages them or forces them to work on the project together towards a common goal, and in the process, a natural change in their relationship is created. To meet the requirements necessary to enroll in this course, participants needed to be able to reach a common goal, trust each other and make decisions as a collaborative learning team [4]. Although there is little empirical evidence on its effectiveness at the college level, developmental social interaction skills through collaborative learning were anticipated. The purpose of this study was to analyze the process of this college class and examine influence robots have in changing the elderly and students' relationship with each other to create new interactions between multi-generations and robots in modern society.

3 Procedure

3.1 Cooperative Active Learning Class

The project of developing a robot application was conducted as a college course between June and December in 2017. To develop the application, humanoid robot, Pepper, developed by Softbank Robotics Co., Ltd. was used. The class consisted of finding the essence of the task, developing an application idea, prototyping, refining, completing the final product, and at the end of the project, an exhibit in the form of a robot application contest (Table 1). Progress of the class focused on group dynamics, emphasizing the revitalization of communication among students and incorporating the five elements that are the necessary conditions for collaborative learning: positive interdependence, face-to-face responsibility Individual Accountability/Personal Responsibility Interpersonal and Small-Group Skills, and Group Processing [6].

The 12-session project was conducted for seven days over a 3-month period. Sessions 1, 7, and 12 focused exclusively on improving the relationship between the participants and the communication gap due partly as a result of the difference in their age and experiences.

Meet Up was a workshop where participants took on the role of the robot, Pepper. The idea of this exercise was to encourage the elderly participants to use their bodies freely to communicate without worrying about their age and their lack of ICT experience. In a 90 min time period, the two groups were divided into five teams for 90 min where they gathered opinions, and subsequently each group gave a presentation followed by a group discussion and feedback.

Speak Up entailed group work that aimed to increase communication between the two generations. Up until now, the elderly had usually held a leadership position at their workplace and in society. Japanese culture is a hierarchical society where the older generation is respected and holds the power. In this project, rather than be the leaders, the elderly played only a minor role and as the students were in charge, they had to follow what the young people told them to do which put them in an uncomfortable situation. In addition, the elderly and young women tend to be regarded as weak and vulnerable in Japan, so it was this uncertainty as to what their proper role should be in this situation that brought on unexpected tension and problems between the two generations. To clear the air and close the communication gap, a 200-minute workshop, where both groups spent from 20 min to an hour reflecting on the situation and talked about how they could narrow the communication gap. Since communicating outside of the classroom was important, both students and the elderly were encouraged to keep in con tact with each other using email and raised the frequency of feedback.

Fix Up was the final reflection stage of the project where both groups could look back on the course, the completed application and the final competitive contest. It was particularly important to have had these three intervening sessions in order to create closer communication between the two groups. The main purpose of this project was to develop an application, but in the process, all the participants realized that through this experience they were able to form useful skills that could be of important practical use in society.

The trial included eight elderly people (five male and three female). The mean age was 72.4 years old) and there were 12 female college students (eight juniors and four elderly people) in one class. The elderly group consisted of those who had no cognitive or physical disorders in communicating and interacting with others. They were recruited using newsletters/flyers that we regularly circulated to those who were interested in our ongoing project dealing with aging. The college students enrolled in this class as a regular academic course.

Table 1. Development flow

Session	Project	Date
Session 1	MEET UP (Intensive group work) Introduction/Workshop: Activation of creativity	17 July
Session 2	Discussion: Finding the essence of task	17 July
Session 3	Learning: Communication based on applied behavioral analysis/IOT and AI	31 July
Session 4	Discussion: Developing application idea/user interview/lectures	31 July
Session 5	Discussion: Developing and designing application idea	11 Sep
Session 6	Rapid prototyping	19 Sep
Session 7	SPEAK UP (Intensive group work) Implementation: User experiment in community work	20 Sep
Session 8	Discussion: Organizing ideas/refine the application	21 Sep
Session 9	Improvement of application using the data through user experiment	25 Sep
Session 10	Competitive presentation: Talk and realize	9 Oct
Session 11	Discussion: Organizing ideas/Refine the application	16 Oct
Session 12	FIX UP (Intensive group work) Feedback and reflection: Finding meaning in experience	23 Oct

3.2   Developed Application

In the application that was developed, the function to improve empathy by refining the 'asking questions' skill to others was programmed. This application expresses and registers the typical problem (cognitive distortion) with plural characters, and solves the problems of the characters with 'asking the questions' and 'sympathy'. As an answer to the 'asking questions', the actual conversation was registered in the application from an interview with a real person. For the conversation with the robot application, we used a mechanism called QnA Maker of Microsoft to learn answers to questions as AI. Moreover, we added the IoT platform to control the lighting of the room and created the asking system of the robot using the face analyzing system of GLORY Co. Ltd. (2) (Figs. 1 and 2).

The concept and process of application development were summarized in the movie; https://youtu.be/U4AqV8EVHRo

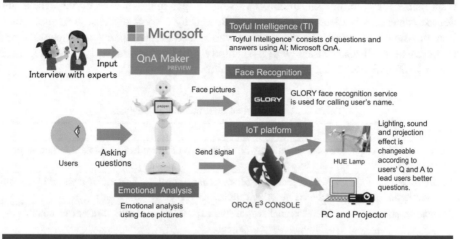

Fig. 1. The system of robot application

Fig. 2. The scene of using the application

4 Method

4.1 Communication Analysis

To analyze the communication change between the elderly and students through the progress of development, the number of utterances in the 10-minute discussions were divided into four categories ('reactive interjections', 'expressive interjections', 'answers with evaluation', and 'questions') (Table 2) which were chosen based on previous studies [7]. Observation was conducted at the beginning of class on the first, third and seventh day of classes for a total of three times (Session 2, Session 8 and Session 11). An educated psychologist and two trained students, majoring in psychology, were responsible for counting the utterances using images and voice recorded with a 360 degree camera. Any discrepancies were followed by a short discussion to reach their final decision. Subjects of the observation who agreed to have their voices recorded and participated in all discussions were five elderly people (three male, two female, mean age = 73.2) and five college students (3 juniors and 2 seniors).

Table 2. Categories of communication analysis

Categories	Detail
Reactive interjections	A representation of acceptance of the other peoples' utterances (at various levels)
Expressive Interjections	A representation of awareness, surprise, disappointment and interest induced by others' utterances and circumstances
Lexical responses	Conventional responsive expressions showing consent to others' opinions, assertions, etc.
Answers/Evaluations	Indication of agreement or disagreement using a short expression (primarily by adjective/adjective verb) in response to the previous person's utterance
Questions	Asking questions when something is unclear

4.2 Measures and Participants

The general information questionnaire covered the usual socio-demographic variables at the pre-test and the satisfaction of participation at post-test. Self-evaluation of their participation consisted of 18 items about ability, ICT literacy, images of robot and relationship with others (Table 2). These items were set based on the survey on ICT education [8] and rated on a five- point Likert scale.

The abilities and skills concerned with competency indicators were evaluated using three instruments, Kiss-18 (Kikuchi's Scale of Social Skills 18 items [8], Time Perspective Scale [10] and Future Attitude Scale [11, 12]. Kiss-18 measures social skills through 18 statements rated on a five-point Likert scale. Kiss-18 provides scores for four basic social skill factors ('conversational skills', 'problem-solving skills', 'work/study skills'). The Time Perspective Scale [10, 12] contains 18 statements evaluated on a five-point Likert scale. The Time Perspective Scale provides scores for 'goal-directedness', 'hopefulness', 'self-fullness', and 'acceptance of past'. As a measure to evaluate fear and anxiety for the future, the Future Attitude Scale was used. Future Attitude Scale 28 statements were rated on a five-point Likert scale. The result is measured as high fear and anxiety as a high score.

The Wilcoxon Signed-Rank Test was applied to compare the pre-test and post-test scores as well as other moderating variables. In addition, the Mann-Whitney U Test was applied to compare the scores of the elderly and students. The participants, six elderly people (three male, three female, mean age = 73.5) and 12 college students (eight juniors and four seniors), completed a survey before and after the course.

Since this study covers the elderly and students, we recruited the participants in writing and verbally presenting the purpose and method of the research. It was possible for any participant to withdraw consent even after agreeing to participate in the research, and students understood their academic grade would not suffer even if they decided not to participate. When we publish our research results, they will receive a consent form with both a verbal and written form indicating that individuals would not be identified.

5   Results

5.1   Communication Change

The numbers of utterances based on the four categories were counted to analyze the communication change between the elderly and students. Figures 3, 4, 5 and 6 indicate the results. The results revealed that the number of students' 'reactive interjections' and 'expressive interjections' were more than elderly people, and these results indicate that the position of the students was passive in the discussion at Session 2. The relationship between the elderly and students was not very good, and they had negative feelings toward the robots. Students reluctantly obeyed the elderly. The elderly expressed their opinions positively. On the other hand, students just passively listened to what the older people said in Session 2.

In order to activate the interaction between the two generations and to decrease the fear of the robot, intensive group work named Meet up was conducted. By acting out using the robot in Meet up, robots became the subjects of joint attention of elderly people and students although they still felt anxious about the robot because they did not feel confident using and controlling the robot. However, students became motivated to create a better relationship with robots and change the triadic interactions from unbalanced status to balanced status (Fig. 7). Students actively tried to take on the role of writing the program and making actions for the robot application. Students' knowledge and technique to control the robot improved from session three to session seven. On the other hand, the elderly's opportunity to improve their skills in controlling the robot decreased. Therefore, intensive group work named Speak Up was conducted. The elderly and students discussed how to solve their complaints in Speak Up.

As the results of observation of communication at Session eight showed, the number of elderly people and students' 'reactive interjections' and 'answers with evaluation' was reversed and the students' numbers were more than elderly people. The number of elderly people and students' 'reactive interjections' and 'answers with evaluation' was reversed and the students' numbers were more than elderly people. The number of elderly people's 'questions' changed slightly, but the number of students' 'questions' increased. Students' feeling towards the robots changed in a positive direction, but the elderly still have fear of robots and they began to put trust in students' ability of control the robots.

In order to close the technology gap, the explanation time about the technique and process was extended. In addition, the length of small group discussions was expanded so that collaborative learning could activate intensively and easily. Through these experiences, participants noticed the importance of teaching each other and asking good questions in order to understand the other generation. Through the group discussions, they decided to develop a robot application to improve the skills of asking to solve the robot's problem. Typical problems in multi-generational interactions were put in the robot's talk. The users of the robot asked the question and solved the problem sympathizing with the robot's feelings. The elderly group and students interviewed each other to clarify the problems in multi-generational interactions. Each generation group could discuss and exchange their opinions frankly by talking objectively for the common goal of robot application creation rather than talking about their troubles directly to each other.

In Session 11, the numbers of both groups were almost the same in all of four categories and the synchronicity of the conversation was promoted. The relationship between the elderly and students became more balanced, and their compassion and intimacy towards the other generation and robots increased.

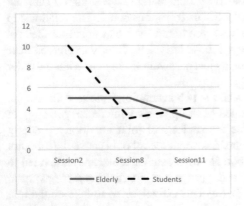

Fig. 3. Number of reactive interjections   **Fig. 4.** Number of expressive interjections

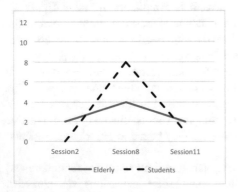

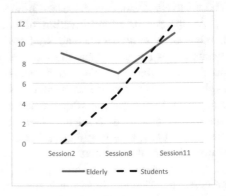

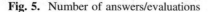

Fig. 5. Number of answers/evaluations   **Fig. 6.** Number of questions

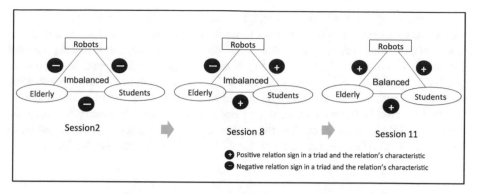

Fig. 7. The change of triadic interactions on three sessions applying Heider's model

5.2 Effects of Participation on Social Skills and Attitude Toward the Future

The differences of the scores between the elderly and college students before participation were analyzed using the Mann-Whitney U Test (Table 3). The result showed elderly people with higher scores than students, and it also revealed that elderly people had higher social skills and a more positive attitude in their time perspective in all subscales of Kiss-18 and Time Perspective Scale. However, the students showed higher scores than the elderly in the Future Attitudinal Scale.

As for the results of the analysis of the difference between elderly people and college students before and after participation, the elderly people' scores were significantly higher ($p < .05$) in 'management skills' and 'job and study skills' on Kiss-18 and elderly people' scores were higher in 'conversation skills' on Kiss-18 ($p < .10$) before participation. However, students' scores in social skills increased to almost the same level of elderly people after participation, and there was no significant difference in social skills after participation in the class.

Regarding the time perspective, although the scores in 'self-fullness' on the Time Perspective Scale was significantly higher ($p < .05$) in elderly people before participation, the scores of 'self-fullness' ($p < .01$) and 'acceptance of past' ($p < .01$) were higher in elderly people than college students after participation. Before class participation, in terms of 'self-fullness" ($p < .01$), the scoring for elderly people was significantly higher. In 'self-fullness' and 'acceptance of past' after class participation, the difference in scores between students and elderly people increased. 'Future Attitude Scale' showed that students' scores were significantly higher ($p < .05$) both before and after class participation, and the students were more uneasy about the future than the elderly.

The results of the analysis, according to the Future Attitude Scale, showed that students' scores were significantly higher ($p < .05$) at both before and after participation, and that students were more anxious about the future than the elderly.

In comparing the scores before and after participation, we analyzed the data of the questionnaire using the Wilcoxon Signed-Rank Test (Table 3). A comparison of the scores of the Future Attitude Scale, before and after participation, showed a signify cant difference ($Z = 1.99$, $p < .05$, $r = -. 81$) in elderly people and showed an improving trend ($Z = -1.82$, $p < .10$, $r = -. 51$), and no significant difference was indicated in the students. In both groups, the score decreased after participation, and it was apparent that participation in the class reduced fear and anxiety toward the future. Although no significant difference was indicated, the students' scores increased after participation in the course in all subscales on both the Kiss-18 and Time Perspective Scales.

Table 3. Scores at before and after participation in the course

	Groups	Pre Mean	(SD)	Median	Mann-Whitney U Test U	p value	Post Mean	(SD)	Median	Mann-Whitney U Test U	p value	Wilcoxon Signed-Rank Test Z	p value	ES r
Kiss-18														
Conversational Skills	Student	16.1	(3.5)	16.0	20.0	0.093 +	16.3	(3.7)	16.0	25.0	0.217 n.s.	-0.620	0.535 n.s.	-.17
	Elderly	18.3	(1.9)	18.5			18.0	(2.1)	17.5			-0.707	0.480 n.s.	-.29
Problem-solving Skills	Student	24.2	(4.5)	24.0	15.5	0.038 *	25.2	(4.5)	25.0	26.5	0.270 n.s.	-1.104	0.270 n.s.	-.31
	Elderly	28.2	(3.3)	27.5			27.7	(3.2)	27.0			-0.756	0.450 n.s.	-.31
Work/Study Skills	Student	12.1	(3.3)	12.0	15.0	0.033 *	12.9	(2.8)	12.0	20.5	0.101 n.s.	-1.314	0.189 n.s.	-.37
	Elderly	14.5	(1.6)	15.0			14.7	(1.2)	14.5			-0.272	0.785 n.s.	-.11
Time Perspective Scale														
Goal Directedness	Student	14.8	(5.1)	13.0	22.0	0.135 n.s.	15.1	(3.9)	16.0	21.5	0.118 n.s.	-0.308	0.758 n.s.	-.09
	Elderly	17.7	(2.4)	18.0			17.3	(1.9)	17.0			-0.816	0.414 n.s.	-.33
Hopefulness	Student	11.1	(3.7)	11.0	21.0	0.113 n.s.	11.8	(3.6)	12.0	31.5	0.505 n.s.	-0.777	0.437 n.s.	-.22
	Elderly	13.5	(1.8)	13.5			12.7	(1.7)	12.0			-0.966	0.334 n.s.	-.40
Self-fullness	Student	16.3	(4.3)	18.0	15.5	0.034 *	16.8	(2.6)	17.0	6.0	0.003 **	-0.358	0.720 n.s.	-.10
	Elderly	20.5	(1.3)	20.0			21.0	(1.5)	20.0			-0.816	0.414 n.s.	-.33
Acceptance of Past	Student	12.5	(3.8)	13.0	25.5	0.233 n.s.	12.8	(3.5)	14.0	15.5	0.037 *	-0.431	0.666 n.s.	-.12
	Elderly	14.3	(2.8)	15.0			16.8	(2.5)	16.0			-0.954	0.340 n.s.	-.39
Future Attitude Scale	Student	85.8	(13)	87.0	10.0	0.010 *	79.9	(17)	83.0	16.0	0.044 *	-1.820	0.069 +	-.51
	Elderly	66.8	(17)	64.5			60.5	(17)	59.5			-1.992	0.046 *	-.81

**$p<.01   *$p<.05   +$p<.10

As for the results of the self-evaluation conducted by the participants after the course (Table 4), the elderly evaluated higher in 'interest in robots' than students. On the other hand, students evaluated higher improvement in such areas as 'vocabulary' and 'ability to think'. Both groups saw a high evaluation with the same generation and multi-generational interpersonal exchanges. The result might have been influenced by the fact that the social skills of the elderly before participation were significantly higher than those of college students, and in addition, college students were often in charge of the technical aspects, such as programming, when working on application development in the groups.

Table 4. Satisfaction of participation

Items	Elderly n = 6		Students n = 12	
	Mean	(SD)	Mean	(SD)
Vocabulary increased	3.83	(0.98)	4.00	(0.74)
Ability to understand the key points improved	3.17	(0.75)	3.67	(0.65)
Ability to take notes of main points improved	2.83	(0.75)	3.75	(0.75)
Ability to think improved	3.50	(0.84	4.33	(0.65)
Knowledge of ICT increased	3.83	(0.75)	3.83	(0.39)
Confidence in ICT increased	2.83	(0.41)	3.42	(0.79)
ICT Literacy improved	2.67	(0.82)	3.33	(0.65)
Interest in robots increased	4.33	(0.82)	4.75	(0.45)
Recognition of the importance of robots increased	3.83	(0.41)	4.42	(0.67)
Confidence using robots increased	3.33	(0.52)	3.75	(0.45)
Relationship with people of the same generation became closer	4.00	(0.63)	4.58	(0.79)
Relationship with people of different generations became closer	4.17	(0.41)	4.75	(0.45)

6 Discussion

In the first trial of the collaborative learning program using a robot with the elderly and college students, during the discussion of the course, an increase of participants' synchronicity was observed and recognized. The results suggest that a balanced interaction between the elderly, college students and robots was formed during the collaborative learning process. This course can be regarded as community of practice since it can be said that the process of becoming a community by sharing knowledge and ideas with those who are of a different generation, exchanging opinions with each other in an equivalent position, could be confirmed in the class. According to Lave and Wenger [2], participating in communities of practice encourage people not only to obtain knowledge and skills but also change the relationship between the environment and others, and the change within their internal environment becomes learning. The class became a place where the people could participate with enthusiasm and interest in the application development and also deepen relationships through mutual exchange of ideas. As for satisfaction of the participants, the data reveals both groups felt a positive improvement in their feelings toward the same generation, the other generation and robots.

Regarding the time perspective, the elderly's 'current fulfillment' and 'past acceptance' scores increased, and the differences of scores with the students increased after the participation. Anxiety and fear of the future in both groups also showed a significant decrease after participation. It was inferred that the elderly had formed a more positive attitude to the 'present' and 'past' by exchanging knowledge and opinions with the younger generation and discovered their possibilities through learning. It is thought that the positive experiences they had being engaged in the community as developers changed the participants' attitude and decreased their anxiety about the future. Taking an independent role as developers helped the participants form a more positive and active attitude toward robots and the triadic interaction between the elderly, young students and robots. The result of the analysis indicated that the class to improve their ability to promote autonomous social interaction. It can be said that a truly inclusive community that accepts the diversity of robots and multi-generations will be realized through sharing common goals and compassionate dialogue.

The limitations of this study include the selection bias of the elderly participants with high competencies and small sample size. It might have been the reason why a significant difference was not observed for many of the scores in both groups before and after the class. To determine the effectiveness of this program, an increase in the number of participants and an expansion of long term studies regarding the effectiveness of the project must be examined further.

Acknowledgments. This study was supported by the Center of Innovation Program from Japan Science and Technology Agency, JST and Individual Research Grants in Doshisha Women's College of Liberal Arts.

References

1. Heider, F.: The Psychology of Interpersonal Relations. Wiley, Hoboken (1958)
2. Lave, J., Wenger, E.: Situated Learning; Legitimate Peripheral Participation. Cambridge University Press, Cambridge (1991)
3. Kusaka, N., Narumoto, J., Tsuchida, N., Masuda, K., Shimomura, A., Katsura, K., Ishikawa, M.: Finding the meaning in life program for older adults: the pathway to wonderful aging. In: The 21th IAGG World Congress of Gerontology and Geriatrics, San Francisco, U.S.A. (2017)
4. Kusaka, N.: The intervention of wonderful aging: a program to find the meaning of life. In: The 31st International Congress of Psychology, Yokohama, Japan (2016)
5. Johnson, D.W., Johnson, R.T., Holubec, E.J.: Circle of Learning; Cooperation in the Classroom, 4th edn. Interaction Book Company, Edina (1993)
6. Johnson, D.W., Johnson, R.T., Smith, K.A.: Active Learning: Cooperation in the College Classroom, 1st edn. Interaction Book Company, Edina (1991)
7. Gardner, R.: When Listeners Talk: Response Tokens and Listener Stance. John Benjamins Publishing Company, Amsterdam (2001)
8. Ogawa, A.: Koreisya eno ICT shiengaku, Tokyo, Japan, Kawashima (2006)
9. Kikuchi, A.: Syakaiteki sukiru o hakaru: Kiss-18 handbook; The social skills are measured, The handbook of Kiss-18, Tokyo, Japan, Kawashima (2007)
10. Shirai, T.: A study on the construction of experiential time perspective scale. Jpn. J. Psychol. **65**(1), 54–60 (1994)
11. Zaleski, Z.: Towards a psychology of the personal future. In: Zaleski, Z. (ed.) Psychology of Future Orientation, pp. 10–20. Towarzystwo Naukowe KUL, Lublin (1994)
12. Shirai, T.: Time orientation and identity in adolescence and middle age. Memoirs of Osaka (1997)

Two-Stage Approach for Long-Term Motivation of Children to Study Robotics

Kateřina Brejchová[1](✉), Jitka Hodná[1], Lucie Halodová[1], Anna Minaeva[1,2], Martin Hlinovský[1], and Tomáš Krajník[1]

[1] Faculty of Electrical Engineering, Czech Technical University in Prague, Prague, Czech Republic
{brejcka1,hodnajit,halodluc,minaeann,hlinovsky,tomas.krajnik}@fel.cvut.cz
[2] Czech Institute of Informatics and Robotics, Czech Technical University in Prague, Prague, Czech Republic

Abstract. While activities aimed to attract the interest of secondary school students in robotics are common, activities designed to promote the interest of younger children are rather sparse. However, younger children from families with parents not working in technical domain have a little chance to be introduced to robotics entertainingly. To fill this gap, we propose a two-stage approach by organizing both programming and technology workshops for children by a volunteering group called "wITches", followed by a robotic competition "Robosoutěž" aimed at children who are already familiar with basic concepts. We describe the proposed approach and investigate the effect of both stages on the number of students, their gender composition and their decisions of the field of study. The gathered data indicate that while the second, robotic competition stage is vital in persuading the children to proceed to study technology and robotics, the first, workshop stage is truly crucial to allow them to enter the field at all. In particular, for more than 70% of the participants, the workshops were the first opportunity to be introduced to robotics.

Keywords: Robotic competitions · Hobby clubs
Secondary school students · Electrical engineering · Younger children
IT workshops

1 Introduction

Nowadays, there are multiple approaches to motivate interest of children in STEM (Science, Technology, Engineering, and Mathematics) all over the world. In the USA [1], robotics engineering is introduced via LEGO robotics from kindergartens to fifth-grade elementary. In Russia [2], the Robotics Center was founded that includes robotic courses for schoolchildren competitions and festivals, camps, courses for teachers and other activities for these purposes. In Austria [3], an approach involving multiple entry points for young people to engage

© Springer Nature Switzerland AG 2019
W. Lepuschitz et al. (Eds.): RiE 2018, AISC 829, pp. 137–148, 2019.
https://doi.org/10.1007/978-3-319-97085-1_14

in the STEM fields was employed. Finally, in the UK [4], a developmental research project between a university and a secondary school to develop extended robotics classes was introduced. Unlike the examples above, there are no long-term approaches to involve younger children in robotics or IT in Czechia.

The primary way to promote interest in studying robotics for secondary school students in Czechia is robotic contests [5,6]. However, in most families, neither father nor mother work in a technical field, and cannot support and motivate their children in participation in robotic competitions. Therefore, along with robotic competitions, it is essential to provide children with a chance to get acquainted with programming and robotics from the very young age. Another important aspect of promoting robotics is making it accessible for beginners with no technical background [7]. While [7] stress the importance of simple, intuitive tools, [8] employ introductory courses prior to participation at a robotic contest.

Surveys have shown that girls who are not exposed to the problematic in a fun way before the age of 15 are less probable to engage in the field later [9]. We hypothesize that this applies to boys as well. Hence, to widen a group of children that will be interested in studying robotics, we employ a two-stage approach, similar to [8] by organizing both programming and technology courses for younger children of ages 10 to 15 by a volunteering group called "wITches" and robotic competitions "Robosoutěž" for elementary and secondary school students at Faculty of Electrical Engineering, Czech Technical University in Prague (FEE CTU) as shown in Fig. 1.

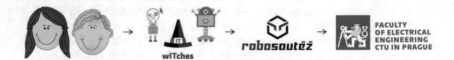

Fig. 1. General outline of the approach

In this paper, we thoroughly describe how both stages work and relate to each other. Furthermore, to investigate how participation in the first stage influences participation in the second one, we analyze the results of questionnaires filled by parents of the participating children. Finally, we provide and analyze a statistics of children that participated in robotic competitions, and whether or not they signed up to one of the study programs at FEE CTU in Prague.

2   Stage I: wITches – Motivating and Educating Children

A volunteering group called wITches consists of female students from FEE CTU in Prague. The group was founded in fall 2015 by four female students after being inspired during the conference womENcourage held in Uppsala, Sweden. New female students were gradually recruited and, finally, in fall 2017 the number of members has grown to 12. The group is funded both by the support of FEE, CTU and by cooperating companies. As a result of the financial support, all organized events can be free of charge. Moreover, building kits and small refreshment are provided on weekend workshops.

The group organizes workshops and hobby classes for children of ages 10 to 15. The events are aimed at beginners and the main purpose is to broaden children horizons in programming, electronics, and robotics. This approach is necessary as most primary school curricula in informatics are based only on the standard handling of proprietary software packages like Microsoft Office and courses on programming or robotics are very rare. Physics is typically taught as a theory only and as a consequence children rarely get the opportunity to acquire hands-on experience in mechanics or electronics. The primary school courses are also aimed at the individuals, rather than collaborative work and project teams tend to be gender homogeneous rather than heterogeneous. The aforementioned aspects deter younger children and especially girls from pursuing their interest in IT and robotics.

First of all, wITches want to show that it is natural that girls are just as excited about programming as boys are. Secondly, we want to introduce children to inter-gender collaboration so that later on, working in gender-heterogeneous teams will be a routine as it is a vital aspect of children future professional life [10]. One of the main advantages of wITches is their age and gender composure: introducing younger children to the world of programming and robotics through hands-on experience by a group of younger female students is somewhat natural. It is also important to mention the significance of the group girls-friendly attitude: graphical materials in pink color, adorable logos of robots and witch hat, friendly and outgoing young female teachers, all of these factors should make girl applicants get rid of the fear to come to workshops or hobby clubs.

Project wITches is intended to carry on even after the founders leave the university. Regular team-buildings for members are organized twice a year, and responsibilities are distributed to other members to ensure the continuation of the group. New members are recruited annually and gradually introduced to the group by following more experienced members and eventually motivated to demonstrate their ideas and initiatives which helps the group to maintain high standards.

2.1   Activities Organized for Children

Group wITches arranges 3–4 h long workshops on weekends, regular hobby clubs during the terms, excursions to industrial partners for children and several other events during the whole year. While workshops are meant as an intensive introduction to the field for a different group of children every time, hobby clubs are regular activities for the same group of participants. Each workshop, as well as regular hobby club, has a capacity of 16 participants, while the capacity of excursions is usually 40 pupils. The teaching activities are aimed at three topics: robotics, programming, and electronics. Nowadays, wITches provide five types of workshops:

1. LEGO robots with LEGO-robotic building kit LEGO Mindstorms NXT,
2. programming in graphical environment Scratch,
3. the basics of electronics with electronic building kit Boffin,
4. electronics with microprocessor Micro:bit,
5. and programming with Ozobot robots focusing on advanced, AI-related algorithms, such as traveling salesman problem.

The LEGO robots and Electronics with Boffin workshops are also held in an advanced version for children that already participated in the introductory workshops. One of the workshops, aimed at the LEGO robots is in details introduced in Sect. 2.2. There is also a short one-minute video [11], which introduces the Ozobot workshop.

At the beginning of the initiative, wITches also organized a workshop on Cutting and editing videos. Until fall 2017, we provided a regular hobby club consisting of three topics: LEGO robots, Electronics with Boffin and Programming in Scratch, but these hobby clubs were not as popular as workshops on the same subject. Programming was firstly taught in graphical environment Alice, but later on we switched to Scratch. Unlike Alice, Scratch programming environment is still supported, being further developed and is more popular among children.

The organized external events for children include excursions to the companies Skype, MSD-IT focused on the symbiosis of IT and the health-care, and Seznam.cz, which runs the most popular Czech search engine and more than 15 other web services. For the general public, we had a stall with robotics-related content on Utubering (festival for children to meet their favorite YouTubers), FEL Fest (a festival organized by FEE), Festival of science and we regularly participate in the Night of scientists (a public event in the Czech Republic). During one of the programs, children had to devise an algorithm, which they later demonstrated in a LEGO robot maze. Another year the program was oriented on Star Wars where visitors could try Ozobot and Micro:bit programming with Princess Leia. We also presented our activities during Open Days of FEE - the first year with building kit Boffin, LEGO Mindstorms NXT and the second year with robots Ozobot and microprocessor Micro:bit. Last year we also organized a competition in programming in Alice and new competition in Micro:bit programming is currently being prepared and planned to commence during June 2018.

As shown in Fig. 2, in two years wITches managed to provide in 31 workshops and 3 hobby clubs for 438 attendants. Among external events, 3 excursions were organized for 67 young participants and some of their interested-enough parents.

2.2 LEGO Robots Workshop

Studies have shown that introducing LEGO workshops in schools improves study results [12,13]. Therefore, LEGO robots workshop was chosen to be taught by our group. To explain the basics of robotics, wITches use LEGO Mindstorms

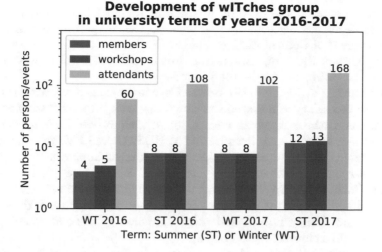

Fig. 2. Development of wITches activity during school terms of 2016–2017.

Education kit that contains a computing unit, sensors, and actuators. The computing unit is a 32-bit microprocessor embedded in the NXT Intelligent Brick that can process inputs from up to 4 sensors, control up to 3 motors and provides 256 KB of FLASH memory and 64 KB of RAM [14]. There are 4 types of sensors that provide data about the robot surroundings: 2 touch sensors, a sensor for detecting intensity and color of (reflected) light, an ultrasonic rangefinder and sound sensor. Finally, the kit contains three motors with built-in reduction gear assemblies and internal optical rotary encoders that provide the motor rotations within one degree of accuracy.

At the beginning of the workshop, wITches describe how robots work using comparison with the human body. For instance, NXT Intelligent Brick is similar to the human brain, it contains programs (comparable to human skills), and every piece (motors, sensors) is connected with the Brick, which is similar to the nervous system in human body. If a human being wants to move, it uses muscles to accomplish that, similarly to a robot which uses motors. To observe surroundings, a robot uses sensors which are analogous to human senses.

In the second part of the workshop, a basic model of the robot is built and then programmed by the workshop participants. The programming tasks are introduced in an increasing level of difficulty: from displaying information using the Intelligent Brick LCD, through sensor-based obstacle avoidance, to maze navigation. At the end of the workshop, the participants compete with each other, demonstrating whose robot drives through the maze the fastest.

3   Stage II: Robotic Competition – Attracting Students

The "Robosoutěž" competition started in 2009 as a final stage of the first term course "Robots", which purpose was to motivate newly accepted students of

"Cybernetics and Robotics" study program at CTU FEE to study more challenging subjects like mathematics and physics. The competition is organized by the Department of Control Engineering FEE CTU. Since FEE and commercial partners support the competition, it is free of charge for all contestants and provides motivating awards for the winners.

Robosoutěž is divided into two categories, one for elementary school students and one for secondary school students competing with university students. Generally, teams of three members have to build their robots for a specific task. The robot can be built only from LEGO MINDSTORMS EV3 or NXT components realizing different functionalities. The task to be implemented is given by the contest rules, which change from year to year. For the past 9 years, the robot challenges included finding a way from a maze, fighting sumo battles, balancing with a ball and carrying it to a finish line through a bumpy road, driving over a river without falling into it, and collecting colored balls and shooting them into the equally colorful targets (Fig. 3).

Secondary school student teams compete in four nomination rounds. Six teams with the best results from each round continue to the final round, where they compete with the best six teams from CTU FEE. Unlike secondary school students, teams from elementary school compete against each other only. The number of participating students during the years is shown in Fig. 7(a).

Fig. 3. Robosoutěž 2017, the task is shooting balls according to colors into targets

There are several similar robotic competitions in the world and in the Czech Republic [6], among the most famous ones are the First Lego League and Robotic Day. Unlike Robotic day, the task to be accomplished in Robosoutěž is published after the participants registration, so that attendants have to solve annually original assignment in a month and a half. Moreover, Robosoutěž is unique in the fact that participants do not need their own LEGO MINDSTORMS set and can borrow them from the contest organizers for free. Hence, everybody can try to make his robot without any financial investments.

4   Evaluation of the Two-Stage Approach

The primary goal of this section is to provide an insight indicating whether our approach meets the intention of attracting younger children to study robotics. In the beginning, we focus on the impact of the first stage which is vital to raise general awareness of the problematics and encourage children to enter robotics. Then, we evaluate the second phase of the process, which should lead in keeping the children interested and possibly to build extensive knowledge and interest in the subject.

We hypothesize that our approach helps to reach a wide range of children and builds interest in broadening their newly acquired knowledge. To evaluate the hypothesis, parents of the children that attended one of the events organized by wITches were asked to fill a questionnaire (see Sect. 4.1 for details).

To evaluate the second phase of the approach, we investigate data from the robotic contest. The hypothesis is that such a competition in robotics strongly motivates students in further interest and engagement in the field, resulting in studying technical fields at universities.

Furthermore, there are more concrete goals that lead to accessibility and popularity of robotics in a wider audience we would like to justify:

1. There are children that come to wITches events without parents with a technical background.
2. There are children coming to wITches with no previous experience in the field.
3. The percentage of girls attending wITches events is larger than the percentage of women at CTU FEE and also larger than the percentage of female students of STEM (Science, Technology, Engineering, and Mathematics) in the Czech Republic.
4. Many children decide to participate in the robotic competition or develop their interest further in some other way after coming to wITches.
5. Participants that attended the robotic competition are often applying to FEE CTU.
6. Popularity of Robosoutěž is growing and attracting a wider range of children.

4.1   Stage I: Questionnaire-Based Feedback

Parents of the children that took part in one of the wITches events answered a questionnaire concerning the following: age, gender, previous experience with robotics before coming to wITches, does one of the parents work in the field, does a child want to continue with robotics/programming in the future and how. The Google Form was e-mailed to all the parents from our database of workshop participants. Parents filled the questionnaire in late 2017 and out of 300 parents whose children attended at least one of the wITches events, 84 filled in the form.

Moreover, the parents left optional feedback, which was always supportive and thankful. Some of the answers also contained advice based on the experience with wITches. Namely, it considered some administration problems, fewer opportunities for experienced children that complete their tasks at workshops much faster than their peers and a request to organize a summer camp for the children. The positive reactions praised level of education that is ensured by having FEE female students as teachers, originality of the project, enthusiasm that children bring with them when they come from the workshop and the fact that all events are free of charge.

As can be seen in Fig. 4, the percentage of female students of STEM fields is very low which is a generally known long-term problem [15]. Group wITches aims to show that women can study and work successfully in STEM fields. The results show that this approach may be a way of attracting girls to IT and robotics. Figure 4 shows that the ratio of female students in STEM was moving around 20–28% in the years 2000–2016 [15]. Notice that these numbers also include other fields than IT and robotics (e.g., architecture, chemistry, etc.). Administration office at FEE CTU suggests that in the past 15 years the ratio of female students at the faculty was fluctuating around 11–13%. In contrast, the ratio of female wITches workshops participants is 34.5%, see Fig. 5. The data were obtained from the conducted questionnaire with 84 respondents. In total, 131 out of 438 workshop participants were girls (29.9%).

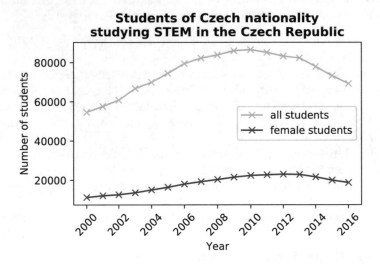

Fig. 4. Female students and all students in STEM in years 2000–2016 (Czech Republic)

Another well-known problem is that children from families with no technical background, no profound knowledge of math or physics, and no extraordinary elementary school teacher, find it very hard to have access to the technical environment in childhood. Since all the first level workshops are designed to be absolutely beginner-friendly, the workshops are often the first contact with robotics

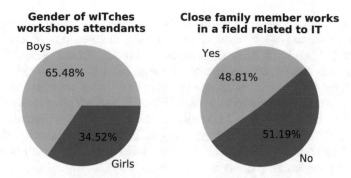

Fig. 5. Gender and background of the wITches workshops attendants

and IT that children experience. The results of the survey in Fig. 6(b) show that 71.4% of attendants had no previous experience in the field at all. The ones that had some experience mostly say that they study robotics on their own using Internet as the main source of information or they have a close relative that is interested in the field himself and teaches the child. Only 7 out of 84 children had previous experience in elementary school education (attended hobby classes organized by their elementary school). Furthermore, approximately half of the attendants have at least one of their parents working in IT- or robotics-related field, see Fig. 5. This means that the influence of the wITches group also reaches beyond the border of the "technical social bubble".

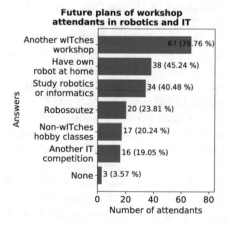

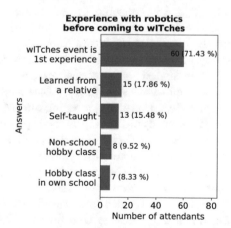

(a) Future plans of wITches workshops attendants

(b) Level of experience participants had before attending wITches workshops

Fig. 6. The impact of wITches workshops

Lastly, we asked the participants whether they plan to develop their newly obtained knowledge further in some way – the results are shown in Fig. 6(a).

Only 3 answers (3.6% of all) were negative, the other considered mostly another wITches workshops (79.8% of all). Another very common option was "to study robotics or IT" and "to have own robot at home". About 24% of workshop participants say that they would like to compete in Robosoutěž in the future. As multiple choice was allowed, it can be evaluated that average attendant whose answer was affirmative, would take part in 2–3 IT and robotics related activities based on her/his experience. In addition, the motivation to partake in other activities could be increased by suggesting suitable hobby classes, websites and education kits necessary for the children to develop further which is something that wITches are working on at the moment.

4.2 Stage II: The Impact of "RoboSoutěž"

Robosoutěž team has been collecting the data about the number of participants and their further activities at FEE, CTU since the competition started in 2009. As shown in Fig. 7(a), the popularity has been continually growing from 4 teams in 2009 to 155 teams which signed up in 2017. It is possible that more teams would have signed up in 2017 if the capacity of the Robosoutěž was sufficient. To suppress this problem, preliminary rounds were introduced in 2013.

One of the most significant competition quality indicator is the number of students that started studying FEE, CTU after participating in the competition. Figure 7(b) shows how many of the former Robosoutěž participants started studying FEE in years 2011–2017 compared to how many students were accepted to FEE in corresponding years (note that y-scales for the curves are different). The data was collected by analyzing the database of Robosoutěž participants and the database of the FEE administration office consisting of students signed up to one of the study programs. In total 132 Robosoutěž participants started studying Faculty of Electrical Engineering during the years 2008–2017. Most of them signed up to study program "Robotics and Cybernetics" (47.97 %), "Open Informatics" (24.2%), "Electronics and Communication" and "Electrical Engineering, Power Engineering and Management" (both 10.6%).

While studying video materials (available in Czech language on-line[1,2]), it has been discovered that the participants seem to be fully absorbed in the problem they solve. Moreover, the participants pointed out some of the problems they faced. Namely, the issues were some constructional or software bugs in the robot that caused its non-deterministic behavior, differences of the terrain on which the robot was trained and tested, which influenced the result of the robot performance and necessity to adjust the robot right at the competition to make it work better. Dealing with such problems prepares contestants for real-life problems where it is necessary to prepare the product for different kinds of possible marginal events.

---

[1] Robosoutěž final, 2017: https://youtu.be/g6IvknPzqxw.
[2] Propagation video, Robosoutěž, 2017: https://youtu.be/DQZ1gp2ZCjY.

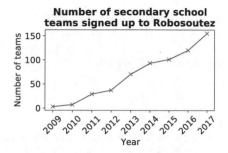

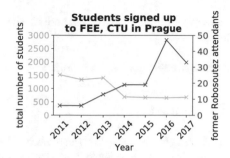

(a) Number of Robosoutěž participants during 2009-2017

(b) Robosoutěž attendants entering FEE vs number of new FEE students

Fig. 7. Robosoutěž popularity and impact

5   Conclusion and Future Work

In this paper, we proposed an approach to reach a broad audience of children and motivating them to involve in activities connected with robotics. The main issues we focused on are a low influence of single events, gender predominance in robotics and the necessity of keeping children interested after the first contact with robotics. We focused on the symbiosis of introductory level workshops organized by young female students (wITches) with more advanced robotic competitions (Robosoutěž).

In Sect. 4.1, we investigated the impact of the proposed two-stage approach. The data indicate that wITches events help to attract a wide range of children since for more than 70 % of workshop participants, wITches events were the first experience in robotic related field.

We also showed that non-negligible percentage of former Robosoutěž contestants decide to study at Faculty of Electrical Engineering. Moreover, we addressed the gender issues and described in detail the proposed solution. We should not make special IT and robotics classes for isolated women and girls; we rather propose inter-gender collaboration as a solution to the problem. In Sects. 2 and 3, we described how both stages of our two-level approach work and how they connect to each other.

Our plans for the future are aimed at increasing the number of kids we influence. To accomplish that, we are preparing new programming contest focused on Micro:bit microprocessor programming, since nobody has organized such a competition in the Czech Republic yet. One of our missions is to connect programming and electronics topics for which the Micro:bit microprocessor is ideal.

While currently, our events take place only in Prague, the capital city of Czechia, in the future we would like to go to other Czech towns with a Road Show. We will contact other universities in Czechia and try to spread our vision. The plan is to visit their town, organize few workshops and, hopefully, to find a group of enthusiastic people, which will start with similar activities. We would like to cooperate closely with the Czech Ministry of Education, Youth and Sports to apply the lessons we learned to the way IT is taught at the primary school level.

Acknowledgements. The work was supported by the CSF project 17-27006Y.

References

1. Cejka, E., Rogers, C., Portsmore, M.: Kindergarten robotics: Using robotics to motivate math, science, and engineering literacy in elementary school. Int. J. Eng. Educ. **22**(4), 711 (2006)
2. Filippov, S., Ten, N., Fradkov, A., Shirokolobov, I.: Robotics education in Saint Petersburg secondary school. In: International Conference on Robotics and Education RiE 2017, pp. 38–49. Springer (2017)
3. Lepuschitz, W., Koppensteiner, G., Merdan, M.: Offering multiple entry-points into STEM for young people. In: Robotics in Education, pp. 41–52. Springer (2017)
4. Samuels, P., Poppa, S.: Developing extended real and virtual robotics enhancement classes with years 10–13. In: Robotics in Education, pp. 69–81. Springer (2017)
5. First: First Lego League. http://www.firstlegoleague.org/. Accessed 18 Jan 2018
6. Robotika.cz: Robotics contests in Czechia. https://robotika.cz/competitions/
7. Cristoforis, P.D., Pedre, S., Nitsche, M., Fischer, T., Pessacg, F., Pietro, C.D.: A behavior-based approach for educational robotics activities. IEEE Trans. Educ. **56**(1), 61–66 (2013). https://doi.org/10.1109/TE.2012.2220359
8. Hernandez-Barrera, A.: Teaching introduction to robotics: Using a blend of problem-and project-based learning approaches. In: IEEE SOUTHEASTCON 2014, pp. 1–5. IEEE (2014)
9. Microsoft: Why Europe's girls aren't studying STEM. https://ncmedia.azureedge.net/ncmedia/2017/03/ms_stem_whitepaper.pdf. Accessed 18 Jan 2018
10. Jiang, M., Li, Y., Zheng, J., Han, X.: J. Comput. Educ. **4**(2), 127 (2017)
11. wITches: Short video about Ozobot workshop by wITches. https://www.youtube.com/watch?v=woBhVurHpw0. Accessed 18 Jan 2018
12. Hussain, S., Lindh, J., Shukur, G.: The effect of LEGO training on pupils' school performance in mathematics, problem solving ability and attitude: Swedish data. J. Educ. Tech. Soc. **9**(3), 182–194 (2006)
13. Lindh, J., Holgersson, T.: Does LEGO training stimulate pupils' ability to solve logical problems? Comput. Educ. **49**(4), 1097–1111 (2007)
14. LEGO: LEGO Mindstorms NXT Education Kit User Guide. https://www.generationrobots.com/media/Lego-Mindstorms-NXT-Education-Kit.pdf. Accessed 31 Jan 2018
15. Ministry of Education of Czechia: University student statisics. http://www.msmt.cz/vzdelavani/skolstvi-v-cr/statistika-skolstvi/data-o-studentech-poprve-zapsanych-a-absolventech-vysokych?lang=1. Accessed 18 Jan 2018

How can we Teach Educational Robotics to Foster 21st Learning Skills through PBL, Arduino and S4A?

Alexandra Sierra Rativa[✉]

Cognitive Science and Artificial Intelligent Department,
Tilburg University, Tilburg, Netherlands
asierrar@uvt.nl, profealexandrasierra@gmail.com

Abstract. This paper provides a framework for (1) how we can foster the 21st century learning skills with educational robotics and some pedagogic strategies, (2) how Problem Based-Learning (PBL) can be used for teaching educational robotics, (3) how we can use a friendly technology to teach educational robotics such as S4A and Arduino, and (4) the evaluation of critical thinking through PBL. Quantitative results are presented to describe frequency codes, co-occurrences and similarity, and linking analysis about students' critical thinking skills during PBL phases. Next to that, the qualitative data provide valuable information on how teachers use educational robotics during PBL, what its advantages and limitations are, and how this topic may develop students' cognitive skills.

Keywords: Educational robotics · 21st Century Learning Skills
Critical thinking · Collaboration · Communication · Creativity
Arduino · S4A · Pedagogic Strategy · Problem Based-Learning · PBL

1 Introduction

There is a growing body of literature that recognizes the importance of robotics in education. Recent evidence suggests that educational robotics can have great potential in classrooms. One of the main obstacles described by Benitti [2] was that the majority of studies about educational robotics were not developed "into classroom activities, i.e., they occur as an after-school program or summer camp program" (p. 5). In contrast, this paper collects the pedagogical strategies used during classroom activities at the high level education in a school. The teaching of educational robotics within the curriculum occurred at formal hours and was integrated with the class of Technology and Art. The project was called Robarte: Robotics, art, and technology. The project was developed during five years in Bogota, Colombia. In particular, the data was used to analyze children's critical thinking (as one of the 21th century Learning Skills). This paper provides an overview of (1) the skills that can be developed through educational robotics

© Springer Nature Switzerland AG 2019
W. Lepuschitz et al. (Eds.): RiE 2018, AISC 829, pp. 149–161, 2019.
https://doi.org/10.1007/978-3-319-97085-1_15

such as 21st century learning skills, (2) the pedagogic strategies that can be used to teach educational robotics, such as Problem Based-Learning, (3) how we can use a friendly technology to teach educational robotics such as S4A and Arduino, and (4) the assessment of critical thinking skills using this pedagogic strategy.

1.1 Educational Robotics for Promoting 21st Century Learning Skills

Teachers' new century challenges are to foster 21st century learning skills in their classroom. These are called the "4Cs": Critical thinking and problem solving, creativity, communication, and collaboration [8,10]. However, an important question is to recognize if educational robotics can develop these types of skills. Extensive research by Benitti [2] of more than seventy articles about educational robotics has shown that educational robotics is focused on fostering skills such as: thinking skills [20], problem-solving skills [11], scientific inquiry skills [22], and social interactions skills [14]. These skills may be associated with 21st century learning skills.

Recent investigations of educational robotics are being directed towards 21st century learning skills. For instance, a study developed by Eguchi [3], who used the concept of programming with LEGO Mindstorms, showed that it is possible to improve collaboration, cooperation, and communication in students. Another example is the international competition called RoboCupJunior. It has focused on project-based learning (PBL) to foster effective competences to solve complex questions and 21st century skills [4]. These high level skills are complex and require a challenging environment to be developed; educational robotics is a powerful tool in this pursuit.

An objective of this study was to foster 21st century learning skills through educational robotics. In this case, it was important that teachers determined a possible definition for each one of these skills. A considerable amount of literature has been published on 21st century learning skills, and yet a generally accepted definition of each ability of the 21st century learning skills is lacking. However, in this study a general concept about these abilities was established. These skills were understood in the classroom in the following way: (1) 'Critical thinking skills' can be comprehended as the capacity of children to study a problem in a reflexive and active way in which they apply knowledge to solve the problem [5,6]. (2) 'Creativity skills' can be recognized as the capacity of children to generate new ideas and to promote their potentiality [22], to develop Mindstorms [16] in novel, original, and innovative ways [9,19]. (3) Communication skills can be understood as children's capacity to write and transmit the ideas and information received in class though technological mediums [1]. Finally (4) 'Collaborative skills' can be understood as children's capacity to work together to resolve a problem with a common objective [23]. Moreover, collaborative skills can also be associated with the decision-making process [24], and interactive social processes (e.g., contribution and participation) [12].

In this specific study it was planned to promote each of the 21st century learning skills. However, this paper just focused on evaluating the ability of critical thinking and problem solving. Despite the importance of every learning skills, each one is very complex and it has sub-skills that can be analyzed in an independent way. Focusing on critical thinking skills, this ability can be divided into subordinate level skills based on Facione [6], who describes six cognitive skills within critical thinking: interpretation, analysis, evaluation, inference, explanation, and self-regulation. Moreover, he described a set of dispositions (attitudes) toward critical thinking such as inquisitive, judicious, truth-seeking, systematic, analytical, open-mind, and confident in reasoning. This study set out to investigate these cognitive sub-skills of critical thinking.

1.2 Some Pedagogic Strategies to Develop 21st Century Learning Skills in Classroom

In this classroom project, pedagogic strategies to foster each learning skill were designed. These strategies were used at a high school level, with interdisciplinary topics and connected with a pedagogical approach. These pedagogic strategies are:

Critical Thinking Skills. The students received a problematic situation about the human body, which they had to solve during the academic year. The characteristics of this problem were: it was related to their real life, the teacher had a possible solution, but the problem had multiple solutions, it was not easy to solve, and it needed the information and technical tools which they received in class to solve. The problem had to do with the sense of sight and touch. The first situation about sight was a person who was light sensitive or experienced photophobia. Students had to think of a way to cure this illness using bionics. The second situation was about a soldier who lost his arm in a war. Students had to ask themselves how they could simulate the missing human arm. Teachers could stimulate critical thinking using brainstorming and mind maps. The mind maps in particular focus on the details of the problem, existing knowledge students may have about the problem, and possible solutions.

Creativity Skills. Students' creativity could be stimulated through art, which can contribute to problem-solving. The mechanical prototypes for the artificial arms and eyes should be related to art concepts. The art concepts included basic drawing, painting and crafts. Another important aspect was the choice of materials used to make the mechanical prototypes, recycled or otherwise. The group of students also chose a theme to design the prototypes, for example Halloween, carnival, gothic, etc.

Communication Skills. The communication skills were stimulated through a continuous writing about the classes given to the students. For this, a digital journal was used. The students wrote about the following questions: What did they learn in class? How did they learn it? Why did they learn it? What did not they learn?, and how will they learn everything that they did not learn? The students answered these question in their journal after each class.

Collaboration Skills. Students were organized in groups, each one composed of four students. Each student had a role which determined their responsibilities in the group. This role was changed several times throughout the academic year so that each participant of the group knew all roles. The roles were as follows: 1. Researcher: This student coordinated the activities during and after class with the group. In addition, the researcher led the presentations of the group and he/she wrote a notebook about their learning process. 2. Secretary: This student carried a folder of every activity during and after class. They also helped the researcher by providing the researcher with scanned photos about the activities performed during class. 3. Lab assistant: This student provided the instruments and material during class. They explained the use of instruments and managed their use by the group in class. 4. Researcher assistant: This student wrote the "Digital Learning Journal" after each class.

1.3   What Is the Problem Based-Learning, PBL?

Teachers require knowledge about pedagogical approaches to foster learning in their students. Technological devices and educational robotics in the classroom without a pedagogical process do not impact learning. Pedagogical approaches give a methodological guide to teachers regarding learning goals, types of skills to foster, topics use in the classroom, their own role, students' role, lesson planning, evaluation methods, pedagogical strategies, and more. Several projects with educational robotics tend to use a constructivist approach. According to Papert and Harel (1991), this constructivism allows students to develop a "learning-by-making" strategy, where the child can learn a "serious" topic in a "playful" way. Problem-Based Learning (PBL) is based on a constructivist approach which is used as a practical pedagogy strategy [17]. PBL allows students to find a solution to a complex problem in a collaborative way [7]. PBL's purpose is for students to problem-solve using an active process. The students must start with a problem which they can connect with their previous experiences. Then, the students learn concepts or information focused on solving the problem. They then have a rich opportunity to apply the knowledge that they received in class. The evaluation of the PBL is associated with their problem-solving and their construction in collaboration. The role of the teacher is as a part of the learning environment, presenting the complex problem and helping to solve the problem with their students. The role of the student is to be active, reflexive, and independent, curious, flexible, and critical. The advantage of PBL is that it can promote higher cognitive skills in the students, foster self-regulated learning, increase motivation, and it can apply technology to real life problems. It is also frequently used in K-12 education [7,21].

1.4   PBL to Teach Educational Robotics

In order to develop an educational strategy to promote 21st century skills through educational robotics, the nine phases of the PBL (Based on Torp and Sage (1998) and Levin (2001)) were redesigned to incorporate a problem with a solution involving robotics and art concepts [13]. The description of these phases are as follows:

(1) **Preparation of the problem:** The students form work groups according to four roles (researcher, secretary, lab assistant and researcher assistant). These roles are developed in an activity called "Caretaker in distress", where the students solve a problem by going through brainstorming about possible solutions using hydraulics and they present the solution in the class. The aim of this phase was for the students to face a small-scale problem and pass, in broad strokes, through all phases of the project.

(2) **Meet the problem:** Students were given a written case that dealt with personal narratives about people who had a visual illness (photophobia), and a disability (amputation of an arm). This case allowed students to become aware of the differences and social exclusion of people with these types of problems, and from there they discussed how they could help in this type of situation.

(3) **Know, need to know, and ideas:** Students made their first mental map, built a problem to address the case and from there wrote what they knew and what they needed to know, and then proposed some solution ideas.

(4) **Problem statement:** The problem was addressed through an essential question and some validity criteria. Students performed an analysis composed of the definition of the problem concept and a writing method to record it.

(5) **Information gathering and sharing:** The students researched in class using the Internet so that they could solve some concerns about new ideas for solutions. Moreover, students were taught in their technology class the use of the S4A program for the programming of the Arduino microcontroller. They could use two types of sensors: photoresistance, which receives signals of light, and the LM35 that measures the temperature. In the arts class, the students saw the theme of color harmonies, which was applied to the construction of human face masks and arm prototypes, as well as the design of the poster presenting their solution to their problem.

(6) **Generate solutions:** The students addressed their problem and proposed more solutions.

(7) **Determine best-fit solutions:** The students analyzed the possible solutions to their problem and chose one.

(8) **Prepare and present solution:** The students built their solution by using circuitry, programing the S4a, and eventually creating a mechanical prototype of eyes and arms. The students presented their solution on Science and Art Day at the school.

(9) **Debrief the problem:** They made a report and final evaluation of their strategies used, the learning they had in this class project and what they still needed to solve.

1.5   Friendly Technology: S4A and Arduino

Educational robotics arises from the need to create learning environments that engage children and young people and stimulate complex high thinking processes. As for robotics, it was born as a science that studies the simulation of the living things, allowing us to glimpse the birth of autonomous machines for the service of human needs. Now, a typical robot system consists of three basic functions: (1) Data acquisition, (2) Execution of the program, and (3) Motion control. The Data acquisition (e.g., through sensors) is understood as the basic function that provides information provided by the sensors of the robot. The data acquisition can directly affect the control of the actuators. The data acquisition of the robot can be composed of sensors of electrical, electronic and mechanical types. (2) The Execution of the program (e.g. human interface) can give us indications of control against the actions performed by the robot. It can be anything from a simple push button in its most primitive form to a complex information system called the "brain" of the machine, such as a microcontroller that is programed by an interface. These microcontrollers can receive the data from the sensors and can send the order to the actuators. Finally, (3) Motion control (e.g., actuators) shows the output of the control system. It can be generated via a sound through a speaker, a movement controller via motor, the temperature of resistance, or the lighting of a led, among others.

One of the main obstacles of robotics is its complexity as a field. Robotics requires that interdisciplinary fields such as math, physics, electronics, programming, mechanical, design, electricity, etc., join together. Specifically, educational robotics is challenging to teachers and students, as they often think that they cannot easily access this knowledge/technology from their classroom. For this to change, educational robotics needs friendly technology and, if possible, lower prices.

Arduino is a low-cost, open-source microcontroller on the market. According to Pan and Zhu (2018), Arduino has the following advantages [15]: (1) it can run on a multiplatform environment (e.g., Windows, Macintosh, and Linux); (2) it is built with an Integrated Development Environment (IDE) (e.g., easy syntax, and graphic user interfaces); (3) it is programmable via USB cable; (4) its soft- and hardware is open-source; (5) it has cheap hardware; (6) it has a dynamic community of users; (7) it was designed in an educational setting.

Arduino can be programmed by C++ in its original version. Moreover, in this project, we used a friendly software to program Arduino. This software is called S4A (http://s4a.cat/). This software can program a microcontroller without the need for the user to know advanced levels of programming. The software is based on Scratch (https://scratch.mit.edu/); however it includes new blocks to control sensors and actuators which can be connected to the Arduino system.

2   Methods

In this paper, the participants were selected in six cases of study depending on their final academic performance. The total sample was selected with 24 students (15 females and 9 males). The students studied in a high school called I.E.D Fabio Lozano Simonelli. This school is located in Bogota, Colombia. The participants were 13 to 16 years old. Of these students, 55% have a computer at home, but 18% of these 55% do not have an internet connection at home, and 45% of students do not have a computer at home. These participants with no computer have a lower socioeconomic status than their peers.

2.1   Apparatus and Materials

Circuit and Programming. Using an electric circuit with Arduino and S4A, we can solve the problem of the disabled person who lost their arm. Based on this circuit, students can simulate tactile sensory experience. We can use a temperature sensor (LM35) to move a servomotor. This servomotor can move the wrist or fingers of the hand. A temperature indicator is a tricolor LED (RGB). The colors of the LED can indicate a high, medium and low temperature.

2.2   Instrument for Data Collection

In order to analyze the data regarding critical thinking skills, the following instrument was designed to collect the information: In the students' digital learning journals, the students registered their daily activities in their Technology and Art classes. They wrote their thoughts and ideas about the guide questions: How they learn and what they learn, how they feel about the proposed activities, and any other data that they can give us regarding the development of their critical thinking skills. This data was collected through an online form on Google. The students had access to the URL of their journal. The students' data was saved automatically in an Excel table.

2.3   Procedure

To conduct this study, we performed a content analysis of the students' journals. This data was collected during one year, consisting of two academic periods. This Journal had 156 complete responses from the class. Specifically, 96 cases were written by girls, and 60 cases by boys. Content analysis is a quantitative method to examine huge text-based data sets, which are qualitative, and classify the occurrence, co-occurrence, and frequency of phrases and keywords which are associated to patterns within the data. The software that was used is QDA Miner version 5. This software allows us to develop a complete analysis of the date obtained.

2.4 Data Treatment

The following steps were used to analyze the data: Step 1: The data from the students' journals was selected from six groups of students. These academic results were provided by members of the school according to the school's standard of evaluation. Participants were separated into two different courses. However, both courses had the same activities during the nine PBL phases. We selected two groups called 1a and 1b which had a superior level (between 5 to 4.5 of the overall score) in their academic results for this class during the academic year. We selected two groups called 2a and 2b, which had a basic academic level (between 3.5 to 4.4 of the overall score) in this class. Finally, we selected two groups called 3a and 3b, which had a low academic level (between 1 to 3.4 of the overall score) in this class. Step 2. The selected data was added to QDA Miner. Step 3: The code of analysis (see Table 1) was designed to have five cognitive sub-skills. These sub-skills were part of the unique category of critical thinking. The table describes the meaning of each code and its associations to words or phrases that can connect to say code. Step 4: The phrases or words written by the student were associated to a code in QDA Miner Software. Step 5: The analysis of the data of each code was carried out. Step 6. Coding by variable was analyzed by phase and group with a bar graph and code occurrence table. Step 7: Coding frequency was analyzed by each code (skills), with a bar graph and code occurrence table. Step 8. Coding co-occurrences were analyzed by each code (skills) with a link analysis graph, co-occurrences table and similarity table.

Table 1. Description of the code of analysis

CODE	Description of the code	Associated words/phrases
Interpretation	Ability to understand the meaning or implication of an extensive variety of activities or experiences	Understand, perceive, I having a clear idea, discovery of the deep meaning of something
Analysis	Ability to identify the relations between declarations, opinions and beliefs	Analysis, connecting the topics taught in class with problem-solving
Evaluation	Ability to evaluate the relations between declarations, opinions and beliefs	Own work is assessed, they evaluated their colleagues in class, considering own grade, the class needs to improve on,
Explanation	Ability to present the results of their own reasoning clearly, coherently, and reflectively	Giving arguments for or against the proposed solution of the problem, explaining topic views during the classes, presenting results
Self-regulation	Ability to self-assess their cognitive experiences, learning processes, and results	Giving a self-assessment about their individual work, expressing aspects to improve, need to improve this skill or activity

3   Results and Discussion

3.1   Coding by Variable: Phases

The first question this educational robotics project aimed to explore was the behavior of the sub-skills of the critical thinking of students during the nine phases of PBL. Moreover, the purpose of this study was discovering how the writing of students can give us a key to understand these types of skills. Table 2 presents the results obtained from the analysis of coding by the 'phases' variable (PBL). This table shows a code of occurrence in the form of row percentage values. The code occurrence was a comparison of the number of cases (complete journal entries) and the codes (sub-skill of critical thinking). For instance, the code occurrence estimates the percentage of occurrence of phrases or words associated to each code. This table shows that the overall percentage of each skill is different in each phase of PBL. Overall, the most occurrences of associated words and topics for all possible skills (e.g. interpretation, analysis, evaluation, explanation, and regulation) were observed in phase 2 (Meet the problem). The second highest occurrence of these words and topics was observed in the phase 1 (Preparation of the problem), phase 3 (Know, Need to Know, and ideas), phase 5 (Information Gathering and sharing), and phase 9 (Debrief the Problem).

Table 2. Code occurrence by phase

Phases of the PBL									
Sub-skills of critical thinking	P1	P2	P3	P4	P5	P6	P7	P8	P9
Interpretation	16.80%	30.80%	14.00%	2.80%	15.90%	3.70%	5.60%	4.70%	5.60%
Analysis	10.30%	26.40%	14.90%	3.40%	21.80%	3.40%	4.60%	6.90%	8.00%
Evaluation	8.30%	30.60%	16.70%	2.80%	22.20%	5.60%	0.00%	2.80%	11.10%
Explanation	14.30%	31.40%	22.90%	0.00%	2.90%	0.00%	0.00%	11.40%	17.10%
Self-regulation	22.10%	26.70%	12.80%	2.30%	14.00%	5.80%	4.70%	3.50%	8.10%

In contrast, the minimum occurrence was observed in phase 4 (Problem Statement), phase 6 (Generate solutions), phase 7 (Determine best-fit solutions), and phase 8 (Prepare and present solution). Surprisingly, in phase 7, no occurrences of words and topics regarding evaluation and explanation appeared in the students' journals; however data regarding interpretation, analysis, and self-regulation skills were found. One interesting finding is that skills behaved differently during long periods of teaching and throughout the learning process. We expected that the percentage of occurrence of each of these skills increased with time. For instance, we expected that the skill 'analysis' in phase 1 would have a much lower percentage of occurrence, while phase 5 would have a medium percentage, and phase 9 would have a high percentage. However, it did not occur. We can infer from the data that each skill depends directly on the activity generated in each class or set of classes. Teachers can improve any critical thinking skill of their students during an educational robotics project. In conclusion, the sub-skills of critical thinking can vary during an academic year, depending on activities performed during class. Moreover, the phases of PBL can influence the percentage of occurrence of each of the sub-skills of critical thinking.

3.2 Coding Frequency: Skills

The second question this educational robotics project aimed to answer was which sub-skills were more frequently mentioned in the students' digital learning journals. The results are analyzed with the distribution of the codes (sub-skill) compared to the number of cases (registers) in the students' journals. From 351 cases in the Journal, the most-mentioned sub-skill of critical thinking was interpretation, followed by analysis and self-regulation. The critical thinking skills less frequently mentioned were evaluation and explanation. One conclusion of my master thesis about this research was that these skills were highly correlated [18]. For instance, the study suggested a strong relationship between explanation, interpretation and analysis skills. On the other hand, it also proposed a relationship of similarity between evaluation and regulation skills. Previous results (in the master thesis) were made by hand without qualitative analysis software. Thanks to an analysis of co-occurring coding though a link analysis graph, in this paper, we can observe the relationships between the abilities. The results show that analysis skills have a strong relationship with interpretation skills (0.22 similarity) and evaluation skills (0.22 similarity). Moreover, evaluation skills have a strong relationship with (0.22 similarity) analysis skills. In conclusion, this study confirms that skills can be associated by co-occurrences. In this paper, contrary to my master thesis, the relationship of similarity was between analysis-interpretation-evaluation skills, as well as interpretation- evaluation skills.

3.3 Coding by Variable: Groups

Finally, the third question this research aimed to answer was how the behavior of the sub-skills of the critical thinking of students compared with their academic results in class. The bar chart 1 shows (Fig. 1) that the maximum occurrence of associated words and topics for all skills (e.g. interpretation, analysis, evaluation, explanation, and regulation) was observed in groups 1a and 1b. The minimum number of occurrences of these was in the group 2a and 3b. Group 3b had a null percentage of occurrence data in evaluation. Group 3a had a higher percentage of occurrence data compared to groups 2a and 2b. One interesting finding is that the groups with better academic results in their overall score frequently mentioned words and topics associated to all sub-skills of critical thinking. Group 1a had a higher academic performance and overall score compared to Group 1b. We can suggest that Group 1a's sub-skills of critical thinking could have given a significant performance advantage compared to other groups'. However, a future study on this subject would be needed to confirm this hypothesis. One unanticipated finding was that Group 3a mentioned more words and topics associated with the sub-skills of the critical thinking than Groups 2a, 2b, and 3b. Group 3a registered a low academic level in this class, and yet this group mentioned more words and topics associated with critical thinking skills compared to groups that obtained a basic level in their academic performance. We do not know which other aspect of the class could have influenced their academic results. These findings suggest that critical thinking can provide a potential advantage for students' academic results. However, it is not the unique determinant factor in their overall score.

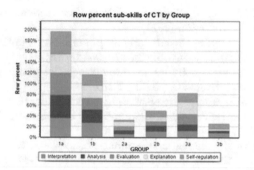

Fig. 1. Graph of the percentage of critical thinking skills by groups.

4   Conclusions

In conclusion, 21st century skills can be fostered through educational robotics. It is recommended that teachers who wish to teach educational robotics make use of a pedagogical strategy. This pedagogical strategy can provide an effective guide to implement new technology in the classroom. In addition, to teach educational robotics, it is recommended to use an easy and friendly technology. In this paper, we suggest a possible activity that you can perform with Arduino and S4A in the classroom. Although it was not possible to assess every one of the 21st century learning skills, this paper had the opportunity to assess critical thinking skills during the nine phases of problem-based learning. An interesting finding is that critical thinking skills behaved differently during long periods of teaching and learning, which depended on the activities that are developed with students. More research is needed to investigate how we can assess 21st century learning skills through quantitative measurement, such as standardization tests. Moreover, we need to learn how we can measure the effectiveness of educational robotics in the fostering of 21st century learning skills. There is abundant room for further progress in educational robotics, such as accessibility and interaction. In future investigations, it might be interesting to use virtual robots in the development of 21st century learning skills.

Acknowledgements. We thanks our research colleagues at the Fabio Lozano Simonelli School. Special thanks to Arturo Hernandez, Sandra Cardenas, Boris Rocha, Fanny Tovar, and Yenny Aldana for data collection and participation in Robarte project. This educational robotics project has available an online video (2018) with more than one million of views in YouTube. This video is called Robarte: de como la robotica le roba estudiantes a la desesperanza (see link: https://www.youtube.com/watch?v=P8b6bUXx50A) in Spanish language.

References

1. Bellanca, J.A.: 21st Century Skills: Rethinking How Students Learn, 1st edn. Solution Tree Press, Bloomington (2011)
2. Benitti, F.B.V.: Exploring the educational potential of robotics in schools: a systematic review. Comput. Educ. **58**(3), 978–988 (2012)
3. Eguchi, A.: Educational robotics for promoting 21st century skills. J. Autom. Mobile Robot. Intell. Syst. **8**(1), 5–11 (2014)
4. Eguchi, A.: RoboCupJunior for promoting STEM education, 21st century skills, and technological advancement through robotics competition. Robot. Auton. Syst. **75**, 692–699 (2016)
5. Ennis, R.H.: Critical thinking: a streamlined conception. In: The Palgrave Handbook of Critical Thinking in Higher Education, pp. 31-47. Palgrave Macmillan, New York (2015)
6. Facione, P.A.: Critical thinking: what it is and why it counts, pp. 1–30 (1998)
7. Ferreira, M.M., Trudel, A.R.: The impact of problem-based learning (PBL) on student attitudes toward science, problem-solving skills, and sense of community in the classroom. J. Classr. Interact. **47**, 23–30 (2012)
8. Germaine, R., Richards, J., Koeller, M., Schubert-Irastorza, C.: Purposeful Use of 21st Century Skills in Higher Education. Publication of National University, pp.19–29 (2016)
9. Gotz, I.L.: On defining creativity. J. Aesthet. Art Crit. **39**(3), 297–301 (1981)
10. Hakkinen, P., Järvelä, S., Mäkitalo-Siegl, K., Ahonen, A., Näykki, P., Valtonen, T.: Preparing teacher-students for twenty-first-century learning practices (PREP 21): a framework for enhancing collaborative problem-solving and strategic learning skills. Teach. Teach. **23**(1), 25–41 (2017)
11. Hussain, S., Lindh, J., Shukur, G.: The effect of LEGO training on pupils' school performance in mathematics, problem solving ability and attitude: Swedish data. J. Educ. Technol. Soc. **9**(3), 182–194 (2006)
12. Kinsella-Meier, M.A., Gala, N.M.: Collaboration: definitions and explorations of an essential partnership. Odyssey: New Dir. Deaf Educ. **17**, 4–9 (2016)
13. Levin, B.B.: Energizing Teacher Education and Professional Development with Problem-based Learning. Association for Supervision and Curriculum Development, Alexandria (2001)
14. Nugent, G., Barker, B., Grandgenett, N., Adamchuk, V.: The use of digital manipulatives in k-12: robotics, GPS/GIS and programming. Paper presented at the IEEE Frontiers in Education Conference, pp. 1–6. IEEE, San Antonio, October 2009
15. Pan, T., Zhu, Y.: Getting started with Arduino. In: Designing Embedded Systems with Arduino, pp. 3–16. Springer, Singapore (2018)
16. Papert, S.: Mindstorm. Basic Book, Inc. Publishers, New York (1980)

17. Papert, S., Harel, I.: Situating constructionism. Constructionism **36**(2), 1–11 (1991)
18. Sierra, A. R. Proyecto Robarte: Promover los procesos de investigación por medio de la robótica y las TICS utilizando un enfoque constructivista en las áreas de tecnología y artes en un colegio distrital. Master in Education Research, University of the Andes, Bogota. http://repositorio.uniandes.edu.co/xmlui/handle/1992/6624. Accessed 7 Feb 2018 (2015)
19. Sternberg, R.J.: The nature of creativity. Creat. Res. J. **18**(1), 87–98 (2006)
20. Sullivan, F.R.: Robotics and science literacy: thinking skills, science process skills and systems understanding. J. Res. Sci. Teach. **45**(3), 373–394 (2008)
21. Torp, L., Sage, S.: Problems as Possibilities: Problem-based Learning for K − 12 Education, 2nd edn. Association for Supervision and Curriculum Development, Alexandria (1998)
22. Williams, W.M., Yang, L.T.: 19 organizational creativity. In: Sternberg, R. (ed.) Handbook of Creativity, pp. 373–390. Cambridge University Press, Cambridge (1999)
23. Winn, J., Blanton, L.: The call for collaboration in teacher education. Focus Except. Child. **38**(2), 1–10 (2005)
24. Wood, D.J., Gray, B.: Toward a comprehensive theory of collaboration. J. Appl. Behav. Sci. **27**(2), 139–162 (1991)

How Does Participation in FIRST LEGO League Robotics Competition Impact Children's Problem-Solving Process?

Xiyan Chen[✉]

Moray House of Education, University of Edinburgh, Edinburgh, Scotland, UK
hcl_ch@outlook.com

Abstract. Educational Robotics constitutes one of the vital resources in the engagement of children especially regarding hands-on applications of STEM. Featuring competitions such as the FIRST LEGO League (FLL) cultivate in children an affinity for and interest in STEM subjects and are integral in honing their problem-solving skills. This paper aims at reporting on a study that investigates the perceptions of children and adults on how participation in FLL competitions impacts the educational development of individuals. The findings show that children's problem-solving processes are firmly embedded in their FLL Challenge experience.

Keywords: FLL challenge · Robot game · Project design · Core values
Problem-solving

1 Introduction

Aiming to promote young people's interest in STEM and relevant fields of study and future careers, one popular international competition called FIRST LEGO League (FLL) germinated with intent of extending the FIRST concept of technology and science to children from 9 to 16 years old to employ real-world context and hands-on experimentation [1]. This type of robotics competition has been perceived as a platform not just to develop children's STEM abilities but also foster their essential life skills through which they can develop their potential to express themselves and render cogent decisions throughout their lives. Fully ensconcing themselves in these competitions ultimately influences their learning results.

Castledine and Chalmers [2] state that children's problem-solving process and skills are integral to their metacognitive values, assessments, and actions, which are all indicators of learning success or failure in authentic situations. When children realize 'what' and 'how' they have learned during these problem-solving processes in FLL, they are subsequently able to internalize, communicate, correlate, and apply these understandings to the problems they face on a quotidian basis. As Varnado [12] asserts, problem-solving is defined as an intentional action in which children pose, examine, and find solutions to scientific problems by inquiring through a panoply of interactive phases: (1) *Identify and define the problem*: learners observe phenomena, identify the problems, externalize or simplify them by communicating and exchanging ideas with

© Springer Nature Switzerland AG 2019
W. Lepuschitz et al. (Eds.): RiE 2018, AISC 829, pp. 162–167, 2019.
https://doi.org/10.1007/978-3-319-97085-1_16

others; (2) *Research and analyse relevant information*: learners probe problems, plan investigations, and test hypotheses by conducting in-depth research and analysing the findings; (3) *Generate and implement solutions to the problem*: learners can use different approaches they have thought about and apply them to solve problems; (4) *Evaluate and revise the best possible solution*: learners reflect on their problem-solving experience and think about if there could have more optimal solutions.

2   FIRST LEGO League Challenge

The challenge of FLL evinces ideas about how to enhance children's STEM skills during a problem-solving process, which includes the following three components: (a) robot game, (b) project design, and (c) core values.

In (a) *robot game*, children are asked to design and construct interactive robots with the programming-oriented and engineering-oriented Lego Mindstorms System and Lego Bricks to delve into active enquiry vis-à-vis the creation of playful experiences [3, 4]. During this process, children are granted the freedom to develop ideas and utilize their imagination, which are major factors that underpin the design of their robot [5]. Moreover, using robotic kits can help educators to identify and develop the learning strategies employed by children when confronted by unfamiliar educational technologies [6], because designing, developing, and evaluating the robot are not only imaginative but also are mathematically rich activities for children.

In (b) *project design*, children will present theme-related solutions to the real-world problem they focus on in a myriad of ways to hone concrete thinking about abstract phenomena, thereby creating meaningful research and sharing them in community [7]. Projects in robotics facilitate the development of creative thinking amongst children because they are in the verge of learning about how to create and apply new ideas in a given perspective, and they are able to view existing situations from a different vantage point. In doing so, these children can articulate alternative explanations and forge new correlations that yield more positive outcomes [8]. Furthermore, children want to learn something interesting and challenging but controllable and appropriate for their age developmental level [9]; as such, the research project could serve as an effective conduit in the implementation of the concepts identified above in the problem-solving development.

In (c) *core values*, children need to explain how they discover something novel and interesting during robot game and project design. Furthermore, they integrate new ideas, skills, and abilities into their everyday lives. Moreover, they have to consider ideas from team members and respect other teams because the spirit of friendly competition is also important. Through such experiences, young participants will be able to reflect on their whole learning process when solving different problems while also developing their self-awareness and confidence. FLL thus promotes embedded learning for children within realistic contexts and social experiences through collaboration, discussion, and argumentation with peers, or receive social interaction and helpful feedback from the adults [7].

3   Research Design

Interpretivism as the theoretical perspective is suitable to explore these various ideas about children's problem-solving process and experience of different people who have already participated in FLL robotics competition, especially in a small group [10]. In this study, the researcher conducted semi-structured interviews with one self-organized team, including five 11–12 years old children, four parents, and two coaches, which qualified for the 2017 Open European Championship (OEC) in Denmark. Interview questions were designed into three different groups for children, coaches, and parents respectively, included 15 questions for each such as "What steps and strategies have your team done to improve the robot design?" "How can you know the team have given some useful solution to the problems?" This method provided children and adults ample opportunities to clarify their opinions, which could lead to a deeper understanding to the research aim. In addition, questions could be explained during the interviews by repeating or rephrasing them to prevent the respondents from misunderstanding the questions. The responses provided from different categories will then be analysed while concurrently taking some of the opinions in summarizing and emphasizing points in accordance with the research aims.

4   Data Analysis

The semi-structured interviews were audio recorded, and then the recordings of the interviews were all transcribed into a more agreeable format while retaining the participant's general modes of expression [11]. Following transcription, the researcher pinpointed the patterns among different groups' responses and the records were fed onto the QSR NVivo 11 software to facilitate data storage, coding, and thematic analysis of the interviews. Interview data was coded referencing directly to the results gathered from the literature. Although different questions retained their own potential codes as a result of being designed based on the literature review, some answers were marked up with different codes according to the interviewees' response in three aspects – "problem-solving process" "programming" in *Robot Game*, "problem-solving process" "social communication" "real life" "theme setting" in *Project Design*, "fun" "independent learning" "reflection" "cooperation" in *Core Values*.

5   Finding and Discussion

The FLL Challenge is not just a competition for children to learn how to program; it also provides a venue for learning the process of solving different kinds of problems creatively within a team environment. With adults' proper guidance, children are granted the opportunity to experience a myriad of challenges and situations related to real-world within a fun learning environment.

5.1 Robot Game

Similar to Barak and Zadok's [9] point of views, the FLL robot game granted children a chance to get involved in programming while concurrently developing their imaginative cognitive processes in a collaborative way. Furthermore, participation in FLL robot game also embedded the idea of Varnado's [12] problem-solving process for children: after identifying the problems they confronted in programming, they read the guidebook for tips, prioritized the mission, or simplified them, then devised a plan to solve them through discussion. If the task could not be conducted smoothly, they would continue to revise the program and try their best to finish as many missions as possible in 2.5 min. Both parents and coaches viewed this kind of discovery-oriented, experiential, and self-directive robotics activities as forcing children to fully apply constructivism in children's problem-solving process, thereby providing them inspiration for how to develop their future careers related to STEM fields.

5.2 Project Design

The children were attracted by the annual theme-setting of the competition because it was more like a real-life project. It was amazing to see that these children can be so creative and do so much at such a young age. The procedure they used to design their project also presented an intriguing problem-solving process: with coaches' or parents' assistance and proper guidance, they finished their research using newspaper articles, network searching, field study, and even contacted some specialists through emails, which provided them with a more trenchant understanding through various angles to define and analyse a problem, and also helped them come up with an "all-around" solution and present it to others. Parents considered that, rather than learning how to solve real-world problems faced by engineers and scientists in today's world, children could learn about how to deal with the basics of numeracy and literacy together with life skills in these endeavors, including making money management, writing emails to the sponsors, and selling their DIY products for the fundraiser. This fundamental idea of FLL of getting some business sponsors to promote their plan to solve a real-world problem, to a degree, also improved children's social communication skill in an unfamiliar environment. Moreover, FLL provided children a platform to learn broadly during the competition: they learned not just about their own project, but also about the projects completed by other teams, which could inspire their thoughts for more significant and informative real-world problems that might otherwise be ignored.

5.3 Core Values

Similar to Eguchi [13] and Benitti's [14] work, when embedded in a natural experiential learning context for children, educational robotics competition helps cultivate an environment that is both interesting and practical, thereby attracting and keeping their interests in reflecting on their experiences in FLL Challenge. With this kind of theme-based learning, a pathway for children to access "real-world" problems is provided by educational robotics and these young people can evaluate, share, and reflect on what they have learned and developed during participation. Furthermore, it is worth noticing that coaches mentioned the importance of cooperation: learning how to solve problems cooperatively as well as being mindful when encountering difficulties are more important to win, although seizing the precious opportunity to appeal to others with their work is also significant in FLL core values. Another valuable finding from the coaches is that posing guiding questions to children instead of telling the team what to do can benefit children's independent learning as well as problem-solving abilities; this kind of teaching method can be applied in school courses setting.

6   Conclusion and Future Implications

This study employed a qualitative method with semi-structured interviews to identify how FLL robotics competition impacts children's problem-solving development during the process in finishing FLL Challenge. Learning from this competition, the implications for future practice could be illustrated in the following two aspects:

- For FLL organisers, it is good for them to continue real-world theme setting for annual competitions, which to some extent also relates to children's future career selection. They could also develop a topic for the fundraiser activities, which to some extent might enhance children's financial management awareness. What's more, further effort might need to be spent upon providing explicit support and guidance to team coaches, and aside from robot game and research design, developing more guidance of what core values mean.
- For school teachers, it may be beneficial to spend more time on programming abilities and getting more familiar with them, then learn more about authentic problem-solving context from FLL. What's more, learning the coaching model for technology instruction might be beneficial as well: use various ways and rich materials rather than textbook to explain the knowledge, encouraged children to use the materials freely at their own initiative. Furthermore, articulation and reflection are instructional strategies that support unstructured or non-routine problem solving. And teachers could also do more observation with children to find out their strength and more communication with other teachers to improve their concept and ways to guide their children in STEM studies.

References

1. Chalmers, C.: Learning with FIRST LEGO league. In: Society for Information Technology and Teacher Education (SITE) Conference, 25–29 March 2013, pp. 5118–5124. Association for the Advancement of Computing in Education (AACE) (2013)
2. Castledine, A.R., Chalmers, C.: LEGO robotics: an authentic problem solving tool? Des. Technol. Educ. 16(3), 19–27 (2011)
3. Bers, M.U., Portsmore, M.: Teaching partnerships: early childhood and engineering children teaching math and science through robotics. J. Sci. Educ. Technol. 14(1), 59–73 (2005)
4. Demetrikopoulos, M.K., Pecore, J.L. (eds.): Interplay of Creativity and Giftedness in Science. Springer, Dordrecht (2015)

5. Alimisis, D.: Educational robotics: Open questions and new challenges. Themes in Science and Technology Education **6**(1), 63–71 (2013)
6. Gaudiello, I., Zibetti, E.: Using control heuristics as a means to explore the educational potential of robotics kits. Themes Sci. Technol. Educ. **6**(1), 15–28 (2013)
7. Mikropoulos, T.A., Bellou, I.: Educational robotics as mindtools. Themes Sci. Technol. Educ. **6**(1), 5–14 (2013)
8. Nutchey, D., Chandra, V.: Developing general capabilities through FLL (2012)
9. Barak, M., Zadok, Y.: Robotics projects and learning concepts in science, technology and problem solving. Int. J. Technol. Des. Educ. **19**(3), 289–307 (2009)
10. Gray, D.E.: Doing Research in the Real World. SAGE, Thousand Oaks (2013)
11. Dearnley, C.: A reflection on the use of semi-structured interviews. Nurse Res. **13**(1), 19–28 (2005)
12. Varnado, T.E.: The effects of a technological problem-solving activity on FIRST LEGO league participants' problem-solving style and performance (Doctoral dissertation) (2005)
13. Eguchi, A.: Educational robotics for elementary school classroom. In: Society for Information Technology & Teacher Education International Conference, pp. 2542–2549. Association for the Advancement of Computing in Education (AACE) (2007)
14. Benitti, F.B.V.: Exploring the educational potential of robotics in schools: a systematic review. Comput. Educ. **58**(3), 978–988 (2012)

Educational Robotics to Support Social Relations at School

Federica Truglio[1,2]([✉]), Davide Marocco[1,2], Orazio Miglino[1,2],
Michela Ponticorvo[1,2], and Franco Rubinacci[1,2]

[1] Department of Humanistic Studies, University of Naples "Federico II", Naples, Italy
trugliofederica.92@gmail.com
[2] Natural and Artificial Cognition Laboratory, University of Naples "Federico II",
Naples, Italy

Abstract. Robotics is a powerful tool to support learning processes and
to promote social relations. It gives the possibility to work in team thus
stimulating the collaboration among students in a class-group. Therefore
also the students who are marginal in the class have the opportunity to
take part in group activities, improving social relationships with other
students. In this paper we will describe the project "Group activities
and social relations at school" proposed to a Middle School. This project
foresees the comparison of three experimental conditions: educational
robotics lab, coding lab with Scratch and control group. In order to
analyze social relations among students, Moreno sociometric test has
been employed. Results indicate robotics lab effectiveness in promoting
relations at school.

1 Introduction

In the last years educational robotics has achieved an important role in the
field of learning technologies. It represents an effective technological tool to support learning processes and to improve social relations [3,7,8,13,15]. Educational
robotics can be considered the natural evolution of coding. Indeed, it allows to
bring the coding in the real world combining programming with tangible models. Therefore, robotics technologies offer several advantages: an higher intrinsic
motivation, a more immediate error handling, the possibility to share and communicate personal opinions, peer-to-peer education, a bigger sensory involvement and an increase in inclusion [1]. Moreover, educational robotics represents
a methodology that gives the possibility to work in a group stimulating collaboration and cooperation among students belonging to a class-group. Hence,
educational robotics labs allow to establish a relation of interdependence among
students, who have, in this way the possibility to achieve a common goal [6,8]. In
particular, by the means of educational robotic labs students have the chance to
coordinate their efforts, to learn to divide their tasks and to complete a job with
an higher motivation, taking into more consideration, at the same time, other
members of the group. As a consequence, also those students who have been
excluded from class-group have the opportunity to take part in group activity

W. Lepuschitz et al. (Eds.): RiE 2018, AISC 829, pp. 168–174, 2019.
https://doi.org/10.1007/978-3-319-97085-1_17

and to improve relationships with other students. Indeed, educational robotics labs can be considered as innovative tools for the acquisition of both educational skills and relational and communicative abilities [2,15]. In this paper, starting from the above mentioned criteria, we would like to understand if educational robotics lab is actually a tool able to support positive social relations among students in a class-group. In order to answer this question, we have proposed the project "Group activities and social relations at school" (from September to November 2017) to a Middle School in Naples. This project has involved two group activities: an educational robotics lab and a coding lab with Scratch. In the next section we will describe this project in more detail.

2   Materials and Method

2.1   Participants

The participants involved in the project are 70 first-year students of a Middle School (32 males and 38 females with an average age of 10, 48 years). First-year students have been chosen because of their weak ties, especially at the beginning of academic year. The students belong to three different classes. The experimental design foresees three conditions, arranged as follows:

1. a class of 23 students has carried out the educational robotics lab;
2. an other class of 24 scholars has dealt with coding lab with Scratch;
3. the last class, consisting of 23 students, has not been involved in any group activity.

This last one was the control group, the base-line to compare group activities and, in particular, educational robotics lab with about the effects on social relations.

2.2   Lego Mindstorms NXT and Scratch

In our project we have used Lego Mindstorms NXT as robotics technology. It is composed of hardware and software (NXT-G) [1,5]. Scratch is a free programming language which gives the students the opportunity to create their own multimedia and interactive games and paths in a simple and intuitive way, by employing images, music and sounds. In particular, this software helps students to develop a creative mind, a systematic reasoning and a collaborative working method [6].

2.3   Sociometric Test

Sociometric test was developed by Jacob L. Moreno in order to examine the structure and interactions of a group [10–12]. This test can be employed in family therapy, in educational system (class-group and teacher training), in urban planning, in business, in the educational summer camps, in military organization and in religious congregation. In particular, in educational context, the sociometric test can be proposed when the following circumstances occur: conflicts among students, the presence of isolated subjects, and the lack of cooperation in working group. In our project, it has been employed to analyze social relations among students of the three classes. Indeed, sociometric test is an efficient method to obtain a detailed layout of interpersonal relations within a group and to stress the social status of each member. This test focuses on two criteria:

1. *affective-relational perspective* which refers to emotional relationship among students and it is based on their psychological affinities;
2. *perspective related to group organization* which aims at achieving a common purpose.

Every criterion consists of four questions which deal with the topic of preference and rejection towards members of their own group. The questions appearing in sociometric test are as follows:

– *Affective-relational perspective*: 1. Write the names and surnames of those classmates who you would like as room-mates during a school trip. You can write as many names as you like.
2. Write the names and surnames of those classmates who you would not want as room-mates during a school trip. You can write as many names as you like.
3. Write the names and surnames of those classmates who, according to you, would like you as their room-mates during a school trip. You can write as many names as you like.
4. Write the names and surnames of those classmates who, according to you, would not want you as their room-mates during a school trip. You can write as many names as you like.
– *Perspective related to group organization*: 1. Write the names and surnames of those classmates who you would gladly make a working group with. You can write as many names as you like.
2. Write the names and surnames of those classmates who you would not gladly make a working group with. You can write as many names as you like.
3. Write the names and surnames of those classmates who, according to you, would gladly make a working group with you. You can write as many names as you like.
4. Write the names and surnames of those classmates who, according to you, would not gladly make a working group with you. You can write as many names as you like.

The graphical representation of sociometric test is defined sociogram. The latter is a lattice consisting of knots and lines. Knots represent members of a group, whereas lines indicate their relations (continuous lines for choices and dotted lines for rejections). Furthermore, each line has arrows that indicate the direction of relationship. Lines with one arrow represent unidirectional relationships, whereas those with two arrows indicate a bidirectional relationship [10–12]. In order to examine sociometric test data, a procedure is employed which consists in the use of a double entry table named sociomatrix. In this table, the names of the group members and the choices and the rejections expressed and received by each member of the group are situated, in alphabetical order, on the axes of abscissas and ordinates. In particular, the choices are indicated with "1", while the rejections with "−1".

2.4   Procedure

Our project consists of two group activities: an educational robotics lab and coding lab with Scratch. We have assigned randomly the classes to the three experimental conditions (educational robotics lab, coding lab with Scratch and no group activities) [15]. On the 25th September (i.e., before the beginning of labs activities) and on 29th November (i.e., at the end of labs activities), the sociometric test has been administered to the students of three classes (pre-test and post-test). Educational robotics lab and coding lab with Scratch activities have been pursued during six weekly meeting, each lasting one or two hours (for a total of 10 hours). In every meeting, students have been divided in different subgroups. The organization of the subgroups was carried out on the basis of the data obtained by sociometric pre-test. The students belonging to educational robotics lab have carried out the following activities: creation of some posters dealing with technology, building and programming of the robot, realization of road itineraries representing the environment where the robot has travelled. On the other hand, the students of coding lab with Scratch put their efforts into the following activities: realization of some posters regarding the topic of technology, invention of a sprite and stage, programming of sprite in a spatial labyrinth, creation of multimedia road itineraries representing the environment where the sprite has travelled. Therefore, the tasks of students of robotics lab and coding lab are quite the same. The only difference consists in the fact that the students of robotics lab have worked with tangible materials (in order to build the robot and to realize road itineraries), while those of coding lab have carried out their activities exclusively with computer.

3   Results

In this section we will briefly describe the analysis of data sociometric test [15]. The comparison between the data obtained through the sociometric pre and post test has enlightened that a change in the relational dynamics of the three class-groups emerged after two months. This change concerns both criteria of the sociometric test. In particular, by comparing sociograms and socio-matrices of pre-test and post-test (within the affective-relational criterion), we have observed that:

– in the group control, the choices have increased by more than 40% and rejections towards two subjects have grown. Furthermore, a student has suffered from more rejections (from 7 to 12, i.e., an increase of more than 70%);

– in the class of educational robotic lab, the number of connections among students has grown enormously. In fact, the total number of choices have approximately doubled (from 126 to 241 choices). In particular, the status of one student has changed from "isolated" to "famous" subject. Indeed, this participant has risen his number of choices (from 1 to 12). The number of choices has increased considerably for all students except for three of them, who have preserved their status of rejected subjects. Eventually, it is interesting to stress the position of a student. Although his cognitive and psychological problems, this subject has taken an active part in educational robotics lab tripling his choices received by other students (from 4 to 13);

– in the class of coding lab with Scratch, the total number of choices has increased by more than 80% (from 107 to 193), but most choices are unidirectional. Two students have preserved their status of rejected subjects. Finally, the situation of a student is changed, i.e., from 5 choices and 8 rejections to 13 choices and 3 rejections.

Furthermore, in order to confirm the effectiveness of robotic lab, we have carried out a statistical analysis, that is a data analysis considering one factor at three levels [3]. We have examined with the help of the software "R" [9] the scores of post-tests of three experimental conditions by means of ANOVA [4]. It turns out a significant difference between the two means of the three experimental groups (p-value $0.00067 < 0.05$) (see Fig. 1) [15]. In order to test the effective source of differences observed through ANOVA, we have used the analytical contrast with Tukey test [14]. By exploiting Turkey test, we observed an important difference between the means of the two experimental groups "robotics-control" (p-value $0.004 < 0.05$) and a small but statistically not significant difference in the groups "robotics-coding" (p-value $0.052 > 0.05$) (see Fig. 1) [15].

ANOVA

	Df	Sum sq	Mean sq	F-value	P-value
Experimental conditions	2	202,3	101,14	8,168	0,00067

Analytical contrast with Tukey test

	t	p-value	
control-coding	-1,7	0,2157	
robotics-coding	2,373	0,052	
robotics-control	4,023	0,0004	→ p-value < 0,05

Fig. 1. At the top there are the results of Anova, while at the bottom there are the results of analytical contrast with Tukey test.

4   Conclusions and Future Directions

The data analysis indicates an important improvement in the social relations among students who have taken part in educational robotics lab, as compared to students of control group and coding lab. This result can be ascribed to the change of learning perspectives and to the way students relate to each other. In fact, students need to look inside objects tangibly and to compare theirselves to their peers by means of group activities, in order to understand their psychological affinities. Furthermore, a small difference between group of educational robotics lab and coding lab with Scratch has emerged. This result might be due to physical and tangible dimension that distinguishes educational robotics from coding. Therefore, we can conclude that educational robotics labs are actually innovative activities to support positive social relations among students. Nonetheless, this research leaves some open questions. Firstly, in the class of educational robotics lab three students have preserved their status of rejected subject. This issue could be examined in a future research project, which should be aimed at verifying if one of the reasons might be represented by the limited duration of our educational robotics lab. Furthermore, our future purposes will consist in comparing the educational robotic lab with an other creative and physical group activity (such as lab of creative arts, lab about recycling and so on). This comparison will allow us to verify if social relations can be improved specifically by running educational robotics lab or if this result is comparable to the effects of other physical group activity [15].

Acknowledgements. This research has been supported by the EU project CODINC (580362-EPP-1-2016-1-IT-EPPKA3-IPI-SCO-IN) from Erasmus+.

References

1. La robotica educativa (Campustore). www.campustore.it/media/wysiwyg/161205_webinar_robotica.pdf
2. Didoni, R.: Il laboratorio di robotica. TD-Tecnologie Didattiche (27), 29–35 (2002)
3. Gabriele, L., Marocco, D., Bertacchini, F., Pantano, P., Bilotta, E.: An educational robotics lab to investigate cognitive strategies and to foster learning in an arts and humanities course degree. Int. J. Online Eng. **13**(4) (2017)
4. Keppel, G., Saufley, W.H., Tokunaga, H., Violani, C.: Disegno sperimentale e analisi dei dati in psicologia. Edises (2001)
5. Klassner, F., Anderson, S.D.: LEGO MindStorms: not just for K − 12 anymore. IEEE Robot. Autom. Mag. **10**(2), 12–18 (2003)
6. Maloney, J., Resnick, M., Rusk, N., Silverman, B., Eastmond, E.: The scratch programming language and environment. ACM Trans. Comput. Educ. (TOCE) **10**(4), 16 (2010)
7. Miglino, O., Lund, H.H., Cardaci, M.: Robotics as an educational tool. J. Interact. Learn. Res. **10**(1), 25 (1999)

8. Miglino, O., Gigliotta, O., Ponticorvo, M., Nolfi, S.: Breedbot: an edutainment robotics system to link digital and real world. In: International Conference on Knowledge-Based and Intelligent Information and Engineering Systems, pp. 74–81. Springer, Heidelberg (2007)
9. Mineo, A.M.: Una guida all'utilizzo dell'ambiente statistico R. www.cran.r-project.org/doc/contrib/Mineo-dispensaR.pdf
10. Moreno, J.L.: Sociogram and sociomatrix. Sociometry **9**, 348–349 (1946)
11. Moreno, J.L. (ed.): Sociometry Experimental Method and the Science of Society. Beacon House, New York (1951)
12. Moreno, J.L. (ed.): Sociometry and the Science of Man. Beacon House, New York (1956)
13. Rubinacci, F., Ponticorvo, M., Gigliotta, O., Miglino, O.: Breeding robots to learn how to rule complex systems. In: Robotics in Education, pp. 137–142. Springer (2017)
14. Siegel, S., Tukey, J.W.: A nonparametric sum of ranks procedure for relative spread in unpaired samples. J. Am. Stat. Assoc. **55**(291), 429–445 (1960)
15. Truglio, F.: Tecnologie dell'apprendimento: la robotica educativa a supporto delle relazioni sociali nei gruppi classe. Master Degree thesis, Universitá di Napoli "Federico II" (2018)

Technologies and Platforms

AMiRo: A Mini Robot as Versatile Teaching Platform

Thomas Schöpping(✉), Timo Korthals, Marc Hesse, and Ulrich Rückert

Cluster of Excellence Cognitive Interaction Technology 'CITEC',
Bielefeld University, Inspiration 1, 33619 Bielefeld, Germany
{tschoepp,tkorthals,mhesse,rueckert}@cit-ec.uni-bielefeld.de

Abstract. Since robots become increasingly ubiquitous and system complexity increases, teaching university students in robotics is essential for modern studies in computer science. This work thus presents the education curriculum around the *Autonomous Mini Robot* (AMiRo) as a solution to this challenge. The goal is to provide insights to the various fields related to robotics and allow students to specialize in a wide range of topics, depending on their interests. Concept as well as platform have been evaluated and the results reveal a generally positive feedback as well as some issues, for which according solutions are proposed.

1 Introduction

Robots become increasingly ubiquitous in all sectors of economy and even made their way into private homes; a trend that is expected to further speed up in the future [5,9]. At the same time, capabilities of platforms are getting ever-more sophisticated and system complexity increases. This progression of robotics demands for highly qualified hardware engineers and software developers, who are well educated and specialized in this field. In recent years, many platforms have been developed to teach young people at universities, in schools and even at home (in form of toys) in system design, programming, control theory and high-level cognitive processing (e.g. computer vision). Popular systems today are the LEGO Mindstorms series[1], Thymio II [16,19], the Khepera series[2] [17], NAO[3], the TurtleBot series[4], e-puck [15], and many more [1,7,12,20].

The platform this work refers to is the *Autonomous Mini Robot* (AMiRo), which was designed to be a versatile, modular system that can be applied to a

---

T. Schöpping and T. Korthals—First two authors contributed equally to lectures and exercises of this research/work.

This research/work was supported by the Cluster of Excellence Cognitive Interaction Technology 'CITEC' (EXC 277) at Bielefeld University, which is funded by the German Research Foundation (DFG).

[1] https://www.lego.com/en-us/mindstorms/.
[2] http://www.k-team.com/mobile-robotics-products/khepera-iv/.
[3] https://www.ald.softbankrobotics.com/en/robots/nao/.
[4] http://www.turtlebot.com/.

© Springer Nature Switzerland AG 2019
W. Lepuschitz et al. (Eds.): RiE 2018, AISC 829, pp. 177–188, 2019.
https://doi.org/10.1007/978-3-319-97085-1_18

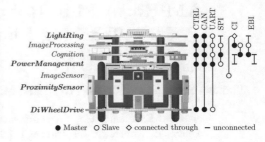

Fig. 1. AMiRo basic setup (left) and uncovered with extension modules attached (right) [7].

Fig. 2. Inner parts of AMiRo and overview of the electrical interfaces with all extensions installed [7].

magnitude of scientific use cases as well as education in university contexts [7,8]. In addition to the physical robot itself, much effort is being spent on the software framework including operating systems, toolchains, interfaces and simulators. As AMiRo is being used in multiple courses, which focus on diverse aspects (not all specifically robotics), the education curriculum is presented in this work.

In Sect. 2 the current AMiRo habitat, including hardware, software and further tools, is described. A selection of courses and projects are presented afterwards in Sect. 3, thereby highlighting the bottom-up approach of the education curriculum. As this concept has been evaluated twofold in a qualitative and a quantitative manner, the results are hence discussed in Sect. 4. Finally, an overall conclusion and future prospect are given in Sect. 5.

2   The AMiRo Habitat

This section briefly describes the AMiRo hardware, its operating systems, software architecture, and compatible tools that are available for the robot. A more detailed description of the system itself is given by Herbrechtsmeier et al. [7].

2.1   Hardware

AMiRo has a circular shape with a diameter of 100 mm at 76 mm height and weights about 500 g. Its handy form factor allows to run tests and small applications directly on the desk right next to a workstation PC, without the requirements of a laboratory environment. One of the robot's primary design concepts is modularity, so the platform can be easily extended and modules can be exchanged. Each module provides a very specific feature set and thus focuses on only a few aspects. Therefore, complexity of each subsystem can be kept reasonably low, whilst the platform is very adaptable and can be applied to a wide range of use cases by customizing the setup. Moreover, this feature allows to quickly setup heterogeneous multi-robot scenarios, by using several AMiRos with differing configurations. Size and appearance, as well as the modular hardware architecture of the system are depicted in Figs. 1 and 2.

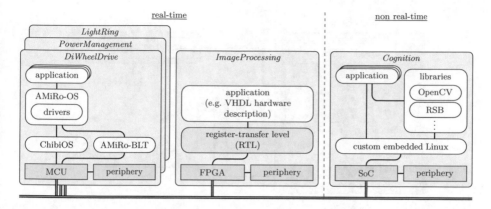

Fig. 3. Schematic system architecture of AMiRo. Whilst the three base modules host MCUs and run a real-time operating system, *Cognition* features a more powerful SoC and provides a full-fledged Linux environment. *ImageProcessing* does not run any OS due to the nature of FPGAs. Modules are linked via several communication interfaces (cf. Fig. 2).

In its basic setup, AMiRo consists of three modules: *DiWheelDrive* facilitates a differential kinematic and is thus responsible for motor control, inertial motion measurement and ground sensing. Charging the batteries, providing power to the system, monitoring those, as well as sensing the near-field environment (via *ProximitySensor*) are tasks of the *PowerManagement* module, which also features a Bluetooth transceiver. Finally, *LightRing* provides LEDs for visualization and interfaces to attach a camera and laser range finder. Each of these modules hosts a microcontroller (MCU) for low-power real-time processing.

In addition to those base modules, two extensions have been developed that provide additional processing power but in two very different ways. Whereas *Cognition* features an SoC (System-on-Chip), *ImageProcessing* hosts an FPGA (Field-Programmable Gate Array). The former is designed for general purpose computation, whilst the latter can be used to (pre)process huge amounts of data (e.g. video streams) efficiently.

2.2   Operating Systems

Since the modules of AMiRo host very heterogeneous processing hardware and focus on various purposes, there are several instances of two different operating systems (OS) running on the platform. Due to the nature of FPGAs, *Image-Processing* is the only module that does not run an instance of either OS. An overview of the software architecture within AMiRo is depicted in Fig. 3.

The MCUs of the base modules are very limited regarding memory size, which makes them incapable of running a Linux-based or comparable OS. They hence run the custom-built real-time operating system AMiRo-OS [21], which is

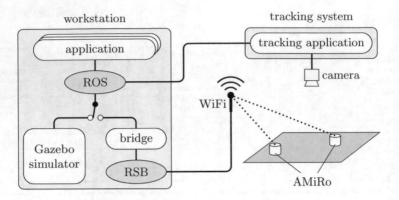

Fig. 4. Illustration of the interaction between applications, simulator, tracking system and AMiRos via the two middlewares ROS and RSB.

based on ChibiOS[5] and extends it by hardware specific configurations, drivers, and interface functions to interact with the underlying bootloader (i.e. AMiRo-BLT[6]). For convenient application development, a C++ abstraction layer is provided.

Being a general purpose module, *Cognition* runs an embedded Linux OS, so a lot of existing software can be executed directly on the robot.

It does not feature real-time capabilities[7], which further increases heterogeneity of the system, another fundamental concept of the AMiRo platform.

2.3   Tools

In order to provide a development environment for the AMiRo platform, several existing tools are used and new ones have been created. On the one hand these tools shall ease many day-to-day tasks like compiling code, interfacing hardware, communication with internal components or external systems, and so forth. On the other hand all software is fully transparent and source code is accessible by developers. This way the initial hurdle is kept on a reasonable level, but students can still acquire profound understanding of how things work together, instead of just using black boxes.

As it is common practice in modern systems to use middlewares in order to abstract the robot's hardware, such are available for AMiRo as well. The most popular system today is ROS (Robot Operating System) [18], which also features a large habitat of useful tools. It is employed for high-level applications that control either an actual robot remotely, or a digital twin in simulation. AMiRo, however, natively runs RSB (Robotics Service Bus) [22], which is more lightweight and offers a more robust communication system. Figure 4 depicts how the two middlewares are combined in order to get the best of both worlds.

---

[5] http://chibios.org/.

[6] AMiRo-BLT is part of the AMiRo-OS project; see https://opensource.cit-ec.de/ projects/amiro-os/ for detailed information.

[7] possible by according configuration of the Linux kernel.

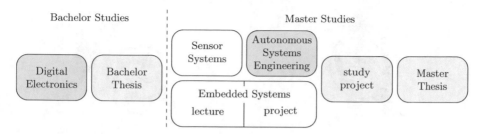

Fig. 5. Excerpt of the education curriculum. AMiRo is used in all courses with dark gray color, whereas white ones do not employ the platform directly. Students are encouraged to work with the system in their theses and study projects.

For running concurrently and reproducable experiments, the Gazebo[8] simulator is used and according models for the robot and its sensors are provided. As shown in Fig. 4, the exact same applications can be either simulated or employed to control actual AMiRos. Moreover, further information of other tools, such as a tracking system, can be accessed as well.

For many experiments the quality of a solution (e.g. precision of navigation or mapping) needs to be quantified. Whilst anything can be measured easily in simulation, this is not true for using real robots. Moreover, some applications require information which can not be acquired by AMiRo itself, such as its precise location. To tackle these challenges, a camera-based tracking tool is used to track multiple robots within the experiment environment setup (cf. Fig. 4).

3 AMiRo in the Education Curriculum

AMiRo is used in multiple annual lectures and projects to teach students in various fields related to robotics. Starting from basic concepts of electrical engineering and computer science, more advanced courses focus on system design of robots and programming applications for such. The overall goal is to introduce students to the various technologies and scientific areas that are involved in modern robotics. Since it is not possible to gain deep knowledge in all those fields, the curriculum provides the possibility to specialize according to interests. In the end, Master students are able to realize hardware modules, develop powerful tools, or implement sophisticated applications for AMiRo within a study project and/or thesis. An excerpt of the education curriculum is illustrated in Fig. 5.

3.1 Teaching the Basics

AMiRo is employed in *Digital Electronics*, which is about the fundamental concepts of modern data processing, starting from transistors to logic circuits, up to complete processors. Specifically for the latter, the FPGA of *ImageProcessing*

---

[8] http://gazebosim.org/.

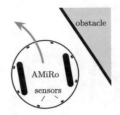

Fig. 6. AMiRo follows a line by reading its four ground facing sensors (top view).

Fig. 7. Using its eight side facing proximity sensors, AMiRo detects obstacles and avoids collisions (top view).

is used to realize a basic but fully functional 32 bit CPU. During additional exercise classes, students develop their own processor core and program it, so that AMiRo is able to drive along a small race track at the end of the term. This task involves not only digital data processing, but also sensor technology, communication and actuator technology.

Two other courses in the curriculum do not employ AMiRo directly, but are closely related to later projects (cf. Fig. 5). *Sensor Systems* is about physical operation of hardware and data fusion techniques on both theoretical and practical level. *Embedded Systems* focuses on hardware design and students have to create a device of their choice in a subsequent project, starting from PCB design, to assembly, up to software development. The skills they gain in these courses are very useful for further lectures and projects with AMiRo.

3.2   Robot Design and Programming

The course that employs AMiRo the most is *Autonomous Systems Engineering*, which comprises system design of robots (hardware and software), implementing fundamental behaviors (i.e. control loops) and realizing complex applications (e.g. navigation & mapping). Following a bottom-up approach, the first topics are again sensor and actuator technology as well as communication interfaces and data processing. The second quarter addresses real-time systems and according requirements for hardware, operating systems and application code. More high-level aspects, such as configuration and building of a custom Linux distribution for an embedded device and diverse communication paradigms (i.a. middlewares), are discussed in the third quarter of the lecture. By then, all aspects of the infrastructure required within robotic systems have been discussed. Hence, the last part of the lecture is split into sensor concepts and control architectures (e.g. subsumption [3]) to solve mapping and navigation tasks.

In addition to the lecture, students gain practical experience in the aforementioned topics by weekly exercise classes using AMiRo. Whilst the first session is about setting up the development environment, getting used to the toolchain and accessing the robot, students are able to solve actual tasks from there on. All software is written in C++, as this programming language is very popular and efficient for embedded systems. The first weeks focus on low-level real-time

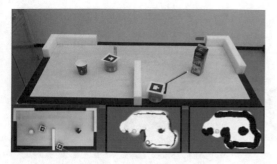

Fig. 8. Cooperative mapping with known poses using AMiRos' proximity sensors.

Fig. 9. Simulation of AMiRo in Gazebo.

applications, thus the basic setup of AMiRo is used. In the second half, exercises employ the ROS-compatible simulation environment to facilitate high-level applications that require relatively large environments. At the end of the term, an additional project task is given, which has to be conducted in groups of two or three within about two and a half months. Several exercise tasks as well as projects are presented in the following.

Basic AMiRo Setup. One of the most basic applications is the line following task. It can be solved by using only *DiWheelDrive* since floor sensors as well as motors reside on this module. The task is to detect a black line on the ground and drive along its path, as depicted in Fig. 6. Students can choose any approach, from fuzzy control to PID controllers, up to neural networks.

Another task is quite similar, but involves communication between *PowerManagement* and *DiWheelDrive*. It is inspired by Braitenberg's vehicles [2] but with a constant forward speed and avoidance of obstacles rather than light sources (cf. Fig. 7). Such are detected by reading the proximity measurements of *ProximitySensor* (cf. Fig. 2).

Cognition and Simulator. First, students shall modify the simulation model of AMiRo and to equip it with a laser range finder so it can precisely measure distances to obstacles. Using this configuration, they have to implement the basic Nearness Diagram Navigation technique by Minguez and Montano [14].

In another task, students shall combine sensor readings with robot positions in order to create a map of the environment, similar to the solution depicted in Fig. 8. This requires to realize an inverse sensor model, which is applied to the Occupancy Grid mapping by Elfes [4]. The results are evaluated by comparison to ground truth as reference.

Project Tasks. Motivated by industrial applications, students shall use multiple robots to replace traditional transportation belts in an assembly line. The basic idea is to use a network of lines on the ground (or on tabletops, respectively) that the robots can use to navigate to target positions within a facility as shown

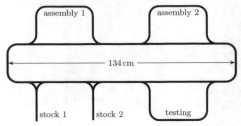
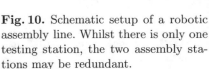

Fig. 10. Schematic setup of a robotic assembly line. Whilst there is only one testing station, the two assembly stations may be redundant.

Fig. 11. Screenshot of a robot soccer implementation. Objects like AMiRos, goals and ball are detected and highlighted in the image.

in Fig. 10. This task can be solved very efficiently with the basic configuration of AMiRo and students can repurpose their solutions of the line following and obstacle avoidance exercises, but it also requires utilization of multiple modules. Students shall furthermore implement a digital twin of the setup in Gazebo (cf. Fig. 9) and solve this task in simulation first before porting the resulting solution to an actual AMiRo. With this approach the benefits and difficulties that arise when using simulators instead of real robots shall be illustrated.

Within another project, students shall make several AMiRos play soccer by using multiple cameras, distinctive ball and goals as well as basic computer vision techniques. This can be achieved by simple blob detection (e.g. red goal, green goal, blue ball) and beacon navigation using the goals as landmarks, as it was done for the solution shown in Fig. 11. Whilst the robots act autonomously, an external system watches the game and acts as referee. Due to the high complexity of this project, it is usually solved by groups of about five students.

3.3 Study Projects and Theses

After students have attended *Autonomous Systems Engineering*, they are well educated in robotics and already got familiar with the AMiRo platform as well as related tools. They are thus able to deal with more sophisticated projects, some of which are presented in the following. Typically each of these projects is conducted within six months (about twelve hours per week).

AMiRo is regularly used in @Home league of the RoboCup competition as part of ToBI (Team of Bielefeld). Multiple robots are employed to act as smart devices and enhance the sensory and actory capabilities of the service platforms Biron and Floka. During a preparation project, a group of several students develop new applications and multi robot demonstrations, which are presented during Open Challenge or Final competitions. In 2015, for instance, several AMiRos performed simultanous localization and mapping plus object recognition (cf. Fig. 8) live on a dinner table, so they could assist a service robot and people by searching for objects and pushing those to desired positions. In another scenario, multiple AMiRos assisted people as mobile beacons to navigate within an apartment in different situations like arrival or emergency evacuation. ToBI eventually even won the world championship in 2016 [13].

During their theses, students are encouraged to engineer or research applications specific for or assisted by AMiRo. Two examplary ones are a Bachelor thesis which resulted in a novel inverse sensor model for proximity sensors applied to mapping, and a Master thesis which investigated the application of semantical occupancy grid maps to informed path planning [10,11]. Furthermore, the development process of AMiRo was accompanied by multiple Bachelor and Master theses, which culminated into a dissertation [6].

4   Experiences and Student Feedback

The quality of the AMiRo platform as well as the whole associated habitat have been evaluated twofold. On the one hand lecturers described their experiences in a qualitative manner, on the other hand a quantitative questionnaire was conducted among students, who have used the system in one or multiple courses. The results of both views are presented in the following.

4.1   Lecturer's View

From a lecturer's perspective, the general impression is that students handle high-level tasks using the simulator or *Cognition* very well. Lower system levels, such as software development for the MCUs of the basic platform and using the according toolchain, seem to be more challenging. It is not possible to decide whether this is caused by the system itself, because of little knowledge about electrical engineering and MCUs, or due to strongly varying C/C++ programming skills. Consequently, motivation and satisfaction of students seem to be better when working within Linux environments and results are generally of higher quality. There are exceptions to this rule, though, so that some people even modify underlying subsystems (e.g. AMiRo-OS) to their advantage, which is interpreted as proof of the benefits of the transparent software architecture. In total, the lecturers impression is that students are able to use the system quite quickly (within few hours), and can also achieve very good solutions to complex tasks at the end of a term or within a project/thesis.

4.2   Student's View

Students who used AMiRo in any course (cf. Fig. 5) have been asked to provide feedback about their experiences with AMiRo via a survey and at the time of writing 29 replies were received. Most participants used the system in 2017 (51%) or 2016 (24%) in the courses *Digital Electronics* (51%) or *Autonomous Systems Engineering* (38%).

First, students were asked about their impressions of AMiRo and the statistical evaluation of their answers is depicted in Fig. 12. Feedback was generally positive and it was found that Master students had a better experience with

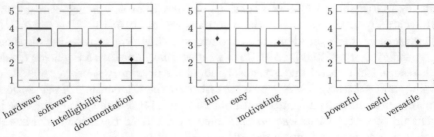

(a) Level of ease regarding hardware and software setup, intelligibilty of the system and quality of documentation.

(b) Feedback about how fun, easy and motivating working with AMiRo was.

(c) Impression about how powerful, useful and versatile AMiRo is.

Fig. 12. Statistical analysis of student feedback about their impressions of AMiRo on a scale from 1 to 5 (higher is better).

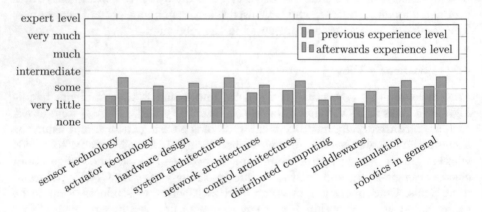

Fig. 13. Mean subjective experience level in various disciplines before and after attending the course.

the system than Bachelor students did. Whilst ratings for the latter are rather mediocre, the former mostly voted positively. Whether this finding is caused by student's level of knowledge or the course they attended can not be determined as these two variables are highly correlated in the results ($\rho = 0.89$). Quality of documentation, however, was rated to be below average consistently by the majority of participants. Although this was intentionally in parts, so that students are forced to analyze and understand the system by themselves, this feedback implies, that the approach did not have the wanted effect.

Students were further asked to rate their experience level in numerous disciplines before and after they attended the course. As depicted in Fig. 13, the mean ratings increase significantly in multiple areas. Again, the answers were found to differ between Bachelor and Master students. Most ratings are around "very little" to "some" for the former, whereas the latter are mostly between "some" and "much". Moreover, mean increase of experience per topic differs between the two groups, a finding which is probably caused by the strong correlation between degree and course, which was mentioned before.

5   Conclusion and Future Prospect

At the beginning of this work, a brief overview of the AMiRo platform and its associated habitat was given. The education curriculum as well as several courses and projects that employ AMiRo have been presented. Finally, the quality of AMiRo as teaching platform was evaluated twofold from a lecturer's and student's point of view, respectively. This revealed a generally positive impression of the platform from both perspectives, but also some issues that need to be addressed in the future.

Aspects of the AMiRo habitat that need to be improved most importantly are the low-level software and overall documentation. Regarding AMiRo-OS, a new major update is planned to be released in the near future, which introduces middleware-based software development for the MCUs. For further development of all components within the AMiRo habitat, extensive documentation will be taken more seriously due to the feedback received.

The education curriculum was found to be beneficial for teaching robotics as well as various related disciplines. In this context, AMiRo has proofed to be a high quality platform for education in university contexts. By its modularity and advanced software and hardware habitat, it is a very versatile platform that can be employed for teaching various aspects of computer science. People are motivated when working with AMiRo and can achieve excellent results in a relative short amount of time. Due to these findings, the platform shall be utilized in further lectures so that students can acquire even more competences in robotics. Furthermore, evaluation of the system and courses will be carried forth in order to further improve the curriculum as well as AMiRo and its habitat. A larger number of participants and optimized survey will help to determine the reasons for several findings of this work, which could not be fully explained yet.

References

1. Betthauser, J., Benavides, D., Schornick, J., O'Hara, N., Patel, J., Cole, J., Lobaton, E.: WolfBot: a distributed mobile sensing platform for research and education. In: Proceedings of the 2014 Zone 1 Conference of the ASEE (2014)
2. Braitenberg, V.: Vehicles – Experiments in Synthetic Psychology, MIT Press (1984)
3. Brooks, R.A.: A robust layered control system for a mobile robot. IEEE J. Robot. Autom. **2**(1), 14–23 (1986)
4. Elfes, A.: Using occupancy grids for mobile robot perception and navigation. Computer **22**(6), 46–57 (1989)
5. Hägele, M.: Robots conquer the world [turning point]. IEEE Robot. Autom. Mag. **23**(1), 118–120 (2016)

6. Herbrechtsmeier, S.: Modell eines agilen Leiterplattenentwurfsprozesses basierend auf der interdisziplinären Entwicklung eines modularen autonomen Miniroboters (2017)
7. Herbrechtsmeier, S., Korthals, T., Schöpping, T., Rückert, U.: AMiRo: a modular & customizable open-source mini robot platform. In: ICSTCC (2016)
8. Herbrechtsmeier, S., Rückert, U., Sitte, J.: AMiRo - autonomous mini robot for research and education. In: Rückert, U., Joaquin, S., Felix, W. (eds.) Proceedings of the 6-th AMiRE Symposium. Springer, Heidelberg (2012)
9. International Federation of Robotics (IFR): Executive Summary World Robotics 2017 Industrial Robots (2017)
10. Korthals, T., Barther, M., Schöpping, T., Herbrechtsmeier, S., Rückert, U.: Occupancy grid mapping with highly uncertain range sensors based on inverse particle filters. In: ICINCO 2016 - Proceedings of the 13th International Conference on Informatics in Control, Automation and Robotics, vol. 2 (2016)
11. Korthals, T., Exner, J., Schöpping, T., Hesse, M.: Semantic occupancy grid mapping framework. In: European Conference on Mobile Robotics. IEEE (2017)
12. López-Rodríguez, F.M., Cuesta, F.: Andruino-A1: low-cost educational mobile robot based on android and arduino. J. Intell. Robot. Syst. Theory Appl. **81**, 63–76 (2015)
13. zu Borgsen, S.M., Korthals, T., Lier, F., Wachsmuth, S.: ToBI – team of bielefeld: enhancing robot behaviors and the role of multi-robotics in RoboCup@Home. In: Lecture Notes on Artificial Intelligence, vol. 9776, pp. 577–588 (2016)
14. Minguez, J., Montano, L.: Nearness diagram navigation (ND): a new real time collision avoidance approach. In: 2000 Proceedings of IEEE/RSJ International Conference on Intelligent Robots and Systems, vol. 3, pp. 2094–2100 (2000)
15. Mondada, F., Bonani, M., Raemy, X., Pugh, J., Cianci, C., Klaptocz, A., Zufferey, J.c., Floreano, D., Martinoli, A.: The e-puck , a robot designed for education in engineering. In: Robotics, vol. 1, no. 1 (2006)
16. Mondada, F., Bonani, M., Riedo, F., Briod, M., Pereyre, L., Rétornaz, P., Magnenat, S.: Bringing robotics into formal education using the thymio open source hardware robot. IEEE Robot. Autom. Mag. **24**, 77–85 (2017)
17. Mondada, F., Franzi, E., Guignard, A.: The development of khepera. In: Experiments with the Mini-Robot Khepera, Proceedings of the First International Khepera Workshop (1999)
18. Quigley, M., Conley, K., Gerkey, B., Faust, J., Foote, T., Leibs, J., Berger, E., Wheeler, R., Mg, A.: ROS: an open-source Robot Operating System. ICRA (2009)
19. Riedo, F., Chevalier, M., Magnenat, S., Mondada, F.: Thymio II, a robot that grows wiser with children. In: Proceedings of IEEE Workshop on Advanced Robotics and its Social Impacts, ARSO (2013)
20. Ruzzenente, M., Koo, M., Nielsen, K., Grespan, L., Fiorini, P.: A review of robotics kits for tertiary education. In: Proceedings of 3rd International Workshop Teaching Robotics (2012)
21. Schöpping, T., Korthals, T., Herbrechtsmeier, S., Chinapirom, T., Abel, R., Barther, M., Kenneweg, T., Braun, C., Rückert, U.: AMiRo-OS (2016). https://opensource.cit-ec.de/projects/amiro-os/
22. Wienke, J., Wrede, S.: A middleware for collaborative research in experimental robotics. In: 2011 IEEE/SICE International Symposium on System Integration (2011)

Teaching with Open-Source Robotic Manipulator

Luka Čehovin Zajc[(✉)], Anže Rezelj, and Danijel Skočaj

Faculty of Computer and Information Science, University of Ljubljana,
Večna pot 113, Ljubljana, Slovenia
{luka.cehovin,anze.rezelj,danijel.skocaj}@fri.uni-lj.si

Abstract. In this paper we present and evaluate the usage of an open-source robotic manipulator platform, that we have developed, in the context of various educational scenarios that we have conducted. The system was tested in multiple diverse learning scenarios, ranging from a summer school for primary-school students, to the course at the university level study. We show that the introduction of the system in the educational process improves the motivation as well as acquired knowledge of the participants.

Keywords: Robotic manipulator · Education · Open-source
Open-hardware · Evaluation

1 Introduction

Robotics is a very attractive and increasingly important research and engineering discipline. Besides a growing commercial interest, it has also a lot of potential in education because of the engagement factor due to the hands-on experience; the students operate a real robots acting in a real world causing real effects in the environment. This is the main reason why robots can play an important role in STEM education in general and motivate students in fields other than robotics itself.

Another property of robotics is that, similar to biological systems, it is very diverse, the most well known categories being robot manipulators, mobile ground robots, and drones. Especially the mobile robots are very popular in education [1–3] because they are reasonably cheap to manufacture and easy to control. Price is a crucial factor in educational robotics, industry-grade robots remain very expensive equipment that only a few educational institutions can afford. This is most clearly shown in the case of robot manipulators that have to be quite complex to ensure a sufficient level of accuracy and reliability. The high price of robot manipulators has resulted in multiple attempts to build a low-cost manipulators [4–8] with varying levels of reliability. These works have focused primarily on the mechanical engineering side of the production such as type and cost of the building material (mostly alloys) and low-level control, leaving a huge

© Springer Nature Switzerland AG 2019
W. Lepuschitz et al. (Eds.): RiE 2018, AISC 829, pp. 189–198, 2019.
https://doi.org/10.1007/978-3-319-97085-1_19

gap in functionality that is required for using the robot in educational scenarios that focus on high-level control, programming and artificial intelligence.

This gap was addressed in our past work [9] where we have presented an open-source robotic manipulator that embraces the recent trend in manufacture of physical objects using the 3D printing technology. Our primary focus in the aforementioned work was a low-cost robot manipulator that is extended into an advanced robotics sensory platform that supports various educational scenarios. In this paper we present our experience with this platform in the context of various educational scenarios that we have conducted in the past year. The system was tested in multiple diverse learning environments, its effects were also evaluated where the evaluation was possible. We show that using the system in the educational process improves the motivation and obtained knowledge of the participants.

In the rest of the paper we first present the platform in Sect. 2, where we summarize its hardware characteristics as well as describe the different programming interfaces that we have implemented. In Sect. 3 we present and discuss the use-cases together with evaluation of their success. We conclude in Sect. 4 with concluding remarks and some ideas for future work.

2   Robotic Manipulator

The main challenge of developing a robotics platform for educational purposes was to design a sufficiently reliable, but still a reasonably low-cost manipulation platform. Our robotic manipulator was initially presented in [9], but we have since then made several improvements towards a more flexible and reliable platform for various educational and research purposes based on our classroom experience. We have also invested a lot of efforts into making the manipulator usable at different skill levels, making it useful for teaching different STEM topics at various levels of education.

2.1   Hardware Platform

The current hardware platform, presented in Fig. 1, is a 5-DOF robot manipulator, based on an improved frame published in [10] and an upgraded open-hardware motor controller, proposed in [11]. The manipulator is supported by a Raspberry Pi 3 computer as the central processing unit and a web camera. We estimate the cost of the entire hardware platform (including 3D printer material, a Raspberry Pi 3 and a USB camera) to approximately 250–300€.

In comparison to the original platform, presented in [9], we have, based on the experience, improved the hardware platform by replacing a simple wooden base for Raspberry Pi 3 computer and a power supply with a custom 3D-printed base and re-designed power supply. This change simplified manipulator setup for users and made the manipulator units easily maintainable allowing easy access to all the components.

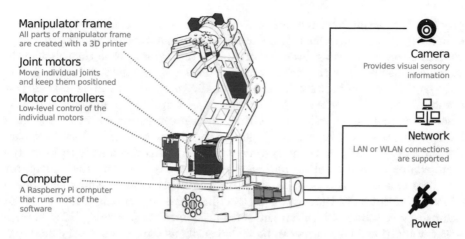

Manipulator frame
All parts of manipulator frame
are created with a 3D printer

Joint motors
Move individual joints
and keep them positioned

Motor controllers
Low-level control of the
individual motors

Computer
A Raspberry Pi computer
that runs most of the
software

Camera
Provides visual sensory
information

Network
LAN or WLAN connections
are supported

Power

Fig. 1. Robotic manipulator and other components of the platform. The base of the robot contains a Raspberry Pi 3 computer and some additional components for power supply. The embedded computer serves as a central computational unit and a bridge between the manipulator and the user.

An important aspect of open-hardware is reproducibility. Since the designs for the manipulator are public, a new unit should be assembled easily by anyone with sufficient knowledge about 3D printing and electronics. Based on our internal evaluation conducted during the manufacture process of ten manipulators for the purpose of our educational activities we have estimated that a skilled engineer can assemble a single unit in less than 24 h. One of the manipulators was even assembled by a keen primary-school student. While this demonstrates the reproducibility principle, we are continuously working on improving this experience in the future by providing written instructions.

2.2 Programming Interfaces

Our system is designed to support multiple programming languages in different interaction scenarios aimed at different skill levels and use-cases. Users can write scripts that are executed directly on the embedded computer of the robotic manipulator or control the manipulator via the HTTP API. All types of interaction are accessible through the web interface running on the embedded computer. The advantage of this approach is that it supports working with the platform without the requirement of having a dedicated client computer with special software installed.

The possibilities of interaction with proposed robotic platform are shown in Fig. 2. The user interacts with the robotic platform via a web-server that supports low-level primitive commands to move the manipulator or access the camera. This type of access is suitable for controlling and monitoring the manipulator directly, which is popular motivational approach for younger students, but can also be used to control the robot from a program executed on a client computer provided that a suitable environment is installed. Alternatively users can submit entire scripts that are executed on the server. This approach is additionally augmented by an in-browser editor and an ability to save scripts on the robot itself, which essentially turns the client computer into a simple dummy terminal that only provides a web browser. This functionality is important in case of shorter events where participants do not have time to set up their computers or in case where their computer skills are limited. In-browser editor currently supports Python scripts as well as Blockly diagrams that are converted to Python within the browser. In both cases the system provides special commands to move the manipulator as well as other hardware-related tasks. We have added support for these two languages because they are frequently used in computer science education. Especially Blockly (as well as its ancestor Scratch) has been used extensively in robotic education and introduction to programming [12–14]. Support for additional languages can be added easily as the system is very modular.

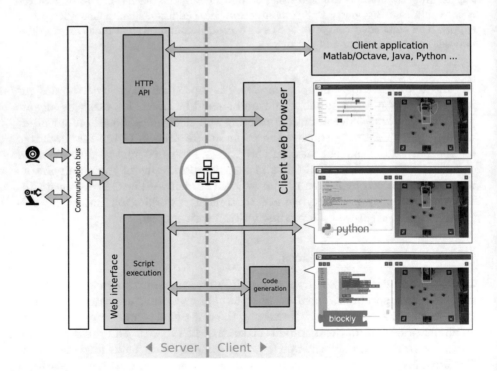

Fig. 2. Programming interfaces overview for the robotics platform. The server side on the left is connected to the client (user) via the local network. The system supports HTTP API for basic manipulator commands as well as execution of entire scripts for more complex use cases.

3   Teaching Use-Cases

The developed platform was so far used in several educational scenarios for the audience of various age and pre-knowledge. In the past two years we have gradually introduced it into curriculum of a course on robotics and computer vision at the university level, we have also organized a summer-school on programming and robotics for primary-school students, as well as use it in educational events that promote the field of computer science. In this section we review these scenarios in more detail and present a motivation for using the developed platform.

3.1   University Course on Robotics and Computer Vision

The initial motivation for development of the described manipulator was a university-level course on robotics and computer vision (RMP). During the course it is important that the students acquire a suitable level of knowledge about both, perception and action part of the robot manipulation. Using the developed robotic platform, we introduce the students to the different topics in a gradual manner. Because the students use the manipulators for the laboratory exercises through the entire semester, we have created a multi-level hierarchy that enables running the same system in three modes. In *simulated environment* the robot manipulator operates in a simulation, while in *augmented environment* the system uses a real camera to perceive the environment but the robot manipulator is simulated and shown superimposed on the top of the camera image. These two environments allow students to deploy the system at home. In the final, *Real-world environment*, the system works with a real manipulator and a camera, using the system that students use in the laboratory.

At the end of the course the students were given an anonymous questionnaire that measured four concepts: (i) how the use of the robotic manipulator influenced their *motivation* for working with robots, (ii) how was it useful to better learn new concepts about *robotics*, (iii) how were they satisfied with the *hardware*, i.e., with the provided platform, and (iv) how useful they found to be the *multi-level* paradigm. Each of the four concepts was measured with four claims, for each of the claims the students provided their attitude on a 5-level Likert scale. Additionally, the students were asked two open questions about the problems with the manipulator and ideas for improvement.

The processed results for the four concepts are summarized in Fig. 3. We can see that the students evaluated the usefulness of the robot manipulator for motivating their work and fostering learning very positively, with an average score of 4.22 and 4.47 respectively, on the scale from 1 to 5. The hardware platform was also positively accepted, with an average score 3.78, although also certain shortcomings of the robot manipulator were exposed. The multi-level paradigm was on average also moderately positively accepted (with an average score of 3.36), however it was not universally accepted by the students, as can be seen in a high variance (1.48) of the answers. This is to some degree understandable as the simulator is not a real substitute for a real manipulator.

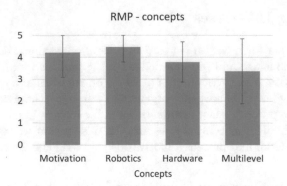

Fig. 3. Results of the survey at the university level course on robotics and computer vision.

We have also analyzed the student surveys of the course for the academic years 2015/16 and 2016/17. The questionnaire is given to the students at the end of the semester by the faculty administration. We are comparing these two years because we have introduced the robotic manipulators in year 2016/17 and the results for year 2017/18 are not available yet. In year 2015/16 9 students provided the answers to the questionnaire, while in the year 2016/17 15 students answered exactly the same questionnaire.

We present here the results of the answers to four questions related to the introduction of the manipulator platform: the *Total* grade, *Overall* satisfaction with the course, encouragement towards independent *Thinking*, and obtained *Competences*. The 5-level Likert scale was used in this case as well.

The results are presented in Fig. 4. We can see that the answers on all four questions improved after the introduction of the robot manipulator in the pedagogical work. The students' positive attitude towards the introduced platform is therefore also reflected in better evaluation of the course.

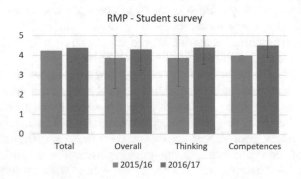

Fig. 4. Comparison of the course grade over years 2015/16 and 2016/17.

3.2   Summer School for Primary-School Level Children

In summer of 2017 we organized a summer school on programming, robotics and computer vision. The summer school was one week long and it was attended by 18 children aged between 12 and 14 years (14 boys and 4 girls). Some children had basic experience in visual programming languages, like Scratch, none of them have worked with robots before. The main goal of the event was to excite children for STEM subjects, and to teach them basic principles of programming (e.g., if-statements, for-loops, etc.), robotics, and computer vision. Because of different backgrounds we have used Blockly as the language of choice. The children were working in pairs, each of them operating a single robot manipulator via the web interface. This way both students were able to solve problems together or individually. Some photos, showing the activity during the summer school, are shown in Fig. 5.

Fig. 5. Photos from the summer school on programming and robotics.

The motivation of children and their attained knowledge was measured with a questionnaire that was given to them before and after the summer school. Motivation was measured with a set of 10 claims, each of them recorded participants attitude towards them on a 5-level Likert scale. The knowledge was measured in terms of self-evaluation with a set of 10 claims. We verified their knowledge also by giving them to solve three assignments related to the basics of programming, robotics and computer vision.

Results of the questionnaire are summarized in Fig. 6. Based on the results we can conclude that the knowledge, both the self-assessed and objective, significantly improved after the summer school. The motivation also improved slightly (the change in all three cases is statistically significant). A reasonably modest improvement in motivation can be attributed to the fact that the students who applied to the summer school were already highly interested in robotics. Several students were also able to program in visual programming languages before the summer school, however they further improved their knowledge (the average percentage of correctly solved programming assignment raised from 61% to 72%). Since the students didn't have almost any pre-knowledge in robotics and computer vision, the improvement in understanding basic concepts in these two research fields is very large (from 22% to 61% and from 11% to 61% respectively).

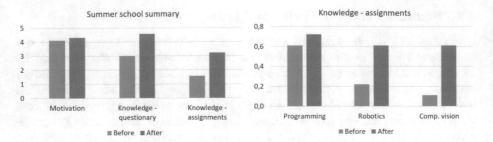

Fig. 6. Results of the questionnaire at summer-school.

3.3   Shorter Educational Events

Our robot platform is also utilized in shorter educational events where one or more robots are used to promote science and robotics. Because the interaction time of visitors is short in this case, their experiences cannot be measured systematically, however, we list the most significant events in this section for completeness together with our findings. In all cases the robots were used to run pre-made Python demos to attract visitors who were able to control the robot manually. Photos from the events are shown in Fig. 7.

Fig. 7. Photos from different short educational events where the robot platform was utilized.

1. **Day of computer science** - In this event several groups of primary school students visited our faculty where multiple short (20–30 min) interactive workshops were held for them. At the workshop the students were introduced to the platform and modified some pre-made Blockly programs. Despite the short turn-around of the groups the system was very stable. Moreover, it sparked a lot of interest in students and in teachers who accompanied them.

2. **The European Researchers' Night** - Our aim at this event was to popularize computer science and robotics research. The visitors (mainly children) were able to see the robotic manipulator at work (performing simple predefined tasks) and also to try to control it. Although they were just weakly supervised by the instructors, they were very quickly able to learn how to operate the robotic manipulator and perform simple tasks. The robot manipulator attracted a lot of attention and brought a lot of fun to the visitors, who also learned something new about the robotics and computer science.
3. *Informativa* **fair** - The fair is aimed at students who are enrolling to universities to help them decide on the field of study. The event was an important test for the platform because it was the first time that the entire event was handled completely by the computer science students who did not have any prior experience with the robot platform and has demonstrated that the setup of the robot is easy enough for someone without an in-depth knowledge about the system's internals.

4 Conclusion

In this paper we have presented our experience with using our open-source robotic manipulator platform at various educational activities that we have organized in the past year. The system was tested in multiple diverse learning environments, from the summer school for primary school students to the university level course. The effects of using the platform were evaluated. We have shown that using the system in the educational process improves the motivation and obtained knowledge of the participants. We have also demonstrated that the platform can indeed be replicated over several units which is a good starting point for an open-hardware project.

We are already in the gradual process of releasing all the plans for hardware as well as the software for the platform so that it may be used by as well as built upon by the community. We are also striving to continuously improve the platform. Based on the results of the evaluations we have also identified several potential improvements. As reported by the students, the accuracy of the robot is insufficient for fine object manipulation tasks. The reason for this are slight differences between the individual units. We are investigating an automatic or semi-automatic manipulator calibration process that would improve the accuracy across all manipulator units. We are also working on improving the simulator that would provide a more realistic testing environment with rigid-body physics simulation. This will hopefully make the simulator a suitable temporary substitute for a real manipulator.

References

1. Yu, J., Han, S.D., Tang, W.N., Rus, D.: A portable, 3D-printing enabled multi-vehicle platform for robotics research and education. In: 2017 IEEE International Conference on Robotics and Automation (ICRA), pp. 1475–1480, May 2017

2. Rezeck, P.A.F., Azpurua, H., Chaimowicz, L.: HeRo: an open platform for robotics research and education. In: 2017 Latin American Robotics Symposium (LARS) and 2017 Brazilian Symposium on Robotics (SBR), pp. 1–6, November 2017

3. Mondada, F., Bonani, M., Riedo, F., Briod, M., Pereyre, L., Retornaz, P., Magnenat, S.: Bringing robotics to formal education: the thymio open-source hardware robot. IEEE Robot. Autom. Mag. **24**(1), 77–85 (2017)

4. Adebola, S.O., Odejobi, O.A., Koya, O.A.: Design and implementation of a locally-sourced robotic ARM. In: 2013 AFRICON (2013)

5. Pronadeep, B., Vishwajit, N.: Low cost shadow function based articulated robotic ARM. In: 2015 International Conference on Energy, Power and Environment: Towards Sustainable Growth (ICEPE) (2015)

6. Pol, M.R.S., Giri, M.S., Ravishankar, M.A., Ghode, M.V.: LabVIEW based four DoF robotic ARM. In: 2016 International Conference on Advances in Computing, Communications and Informatics (ICACCI), pp. 1791–1798 (2016)

7. Karmoker, S., Polash, M.M.H., Hossan, K.M.Z.: Design of a low cost PC interface six DOF robotic arm utilizing recycled materials. In: 2014 International Conference on Electrical Engineering and Information and Communication Technology (2014)

8. Cocota, J.A.N., Fujita, H.S., da Silva, I.J.: A low-cost robot manipulator for education. In: 2012 Technologies Applied to Electronics Teaching (TAEE) (2012)

9. Čehovin Zajc, L., Rezelj, A., Skočaj, D.: Open-source robotic manipulator and sensory platform. In: Lepuschitz, W., Merdan, M., Koppensteiner, G., Balogh, R., Obdržálek, D. (eds.) Robotics in Education, pp. 250–256. Springer, Cham (2017)

10. Gómez, A.C.: Diseño y puesta en funcionamineto de un brazo robótico imprimible (2012)

11. Carter, B.: OpenServo (2008). http://www.openservo.com/

12. Passault, G., Rouxel, Q., Petit, F., Ly, O.: Metabot: a low-cost legged robotics platform for education. In: 2016 International Conference on Autonomous Robot Systems and Competitions (ICARSC), pp. 283–287, May 2016

13. Plaza, P., Sancristobal, E., Carro, G., Castro, M.: Home-made robotic education, a new way to explore. In: 2017 IEEE Global Engineering Education Conference (EDUCON), pp. 132–136, April 2017

14. Koza, C., Wolff, M., Frank, D., Lepuschitz, W., Koppensteiner, G.: Architectural overview and hedgehog in use. In: International Conference on Robotics and Education RiE 2017, pp. 238–249. Springer, Heidelberg (2017)

Prototyping and Programming a Multipurpose Educational Mobile Robot - NaSSIE

Vítor H. Pinto[1,2]([✉]), João M. Monteiro[1,2], José Gonçalves[2,3], and Paulo Costa[1,2]

[1] INESC TEC - Institute for Systems and Computer Engineering, Technology and Science, Porto, Portugal
{vitor.h.pinto,joao.m.monteiro}@inesctec.pt
[2] FEUP - Faculty of Engineering, University of Porto, Porto, Portugal
paco@fe.up.pt
[3] Department of Electrical Engineering, IPB - Polytechnic Institute of Bragança, Bragança, Portugal
goncalves@ipb.pt

Abstract. NaSSIE - Navigation and Sensing Skills in Engineering is a platform developed with the intent of facilitating the acquisition of some skills by Engineering Students, which is a core part of the process of controlling a mobile robot. In this paper, the chosen hardware and consequent physical construction of the prototype as well as vehicle's associated software will be presented. As a use case, this platform was tested during the Robotic Day 2017 in Czech Republic. Preliminary results will also be presented of this year's preparation for the Micromouse competition.

Keywords: Educational robots · Mobile Robotics · Navigation Sensoring · Robotic competitions · Programming

1 Introduction

In recent years, there has been a considerable increase in the use of autonomous robotic platforms. From autonomous vacuum cleaners [7], to robots that carry the necessary materials to the operators in the industry [4], or even military robots that can identify landmines autonomously [5]. All of them have in common the use of sensors, at several levels, as well as control algorithms.

Although the need to use this type of technology is clear for everyone, it is not immediate to develop. In fact, one of the main difficulties when introduced to the Mobile Robotics field is the obstacle detection and navigation. In order to achieve this, it is necessary for the students that will use the robot to acquire knowledge in those fields.

Educational robots provide a platform that can be used as a tool to introduce new concepts and skills to students [6]. Knowing this, the NaSSIE Platform was

© Springer Nature Switzerland AG 2019
W. Lepuschitz et al. (Eds.): RiE 2018, AISC 829, pp. 199–206, 2019.
https://doi.org/10.1007/978-3-319-97085-1_20

created, in order to allow Engineering Students to develop skills in navigation and sensing, through application of learned concepts directly in this robot. As a real life case, to understand if the platform is robust and well designed, it was put to test in the international Robotic Day 2017 competition, and the results are also presented in this paper. Also, this platform was built to serve as a base to participate in Micromouse competitions and some preliminary results will be presented.

Throughout this paper, this platform will be presented: in Sect. 2 it will be presented the design considerations; Sect. 3 describes the implementation, including the mechanical, hardware and software designs. Section 4 introduces the final prototype, and in Sect. 5 some of the possible applications for the robot are stated. Finally, Sect. 6 exhibits the conclusions and the future work.

2   Design Considerations

Since this intents to be an educational platform, robotic competitions can be used as a motivation to introduce problems found in the robotics area. However, these competitions introduce constraints to the robot's design.

2.1   Competitions

The competitions where the robot will participate are Robotic Day 2017 competition, on the RoboCarts contest [3] and on the Micromouse Portuguese Contest [2].

RoboCarts contest consists in a race of up to five fully autonomous robots around the racetrack shown in Fig. 1. All the robots compete at the same time and gain points based on their order at the finish line.

The Micromouse competition is an event where small robot solves a 16×16 maze of $18\,\text{cm} \times 18\,\text{cm}$ Fig. 2. The robot starts int the lower left corner cell and the goal is to reach the four central cells.

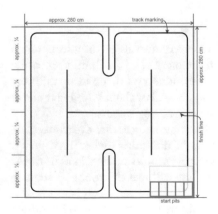

Fig. 1. RoboCarts racetrack [3]

Fig. 2. Micromouse maze [1]

2.2   Design Constraints

Firstly, it is important to note that a platform such as the one proposed, should take into account that its purpose is to help students during their learning process. Also, since there are two competitions where this robot will be used, this also define a set of constraints in the robot design. Given these, a set of specifications are listed below:

- The robot should only contain components that can be acquired easily, so that anyone can replicate it;
- The robot's traction must be differential, since it is, mechanically speaking, the simplest and the more useful to the competitions in which the robot will be used;
- To ease the development by beginner programmers, the processing unit of the platform should have a large community and have a wide availability of resources, such as libraries and tutorials;
- The sensors to be incorporated in the robot must be able to measure distances to the obstacles and walls;
- The robot must be designed to be affordable;
- The robot must be easily mounted and unmounted, for part replacement purposes;

Having decided the material used to build the vehicle, it was necessary to decide what shape would be chosen. Even though several forms have been taken into account for the platform, the rectangular shape was selected. After the shape of the platforms was decided, the front was rounded so that the robot can gain some margin of maneuver in the rotation, thus giving it a greater tolerance to collisions.

3   NaSSIE Design

In the next subsections, all the decisions that were taken in terms of mechanical, hardware and software design are presented.

3.1   Mechanical

The robot must be built with a material that is both rigid and also allows changes during its construction process, because it can be necessary to increase the number of sensors or change the platform design, in which case the material must be easy to process. The final choice was to use extruded PVC plates to construct the body of the vehicle. This material is very easy to process, since it is slightly moldable, which checks the need for cushion on impact; it is also very lightweight and not very expensive. Thus, the vehicle was constructed using two bases of similar size and shape, one lower that will serve as the basis to support the motors and the wheels, as well as the batteries. The upper one support all the other hardware. These were coupled using holders. All the hardware was bolted to the platforms, with the exception of the batteries.

3.2 Electronic Hardware

Regarding the hardware and considering the specification that the hardware should be easily acquired, the robot was built only using simple components. As a power supply for the whole system, two Lythium-Polymer Batteries were used. These batteries are connected in series. There is a power button that allows to turn the robot on and off. The batteries power the motor driver directly, while a step-down regulator lowers the voltage to power the rest of the system: a microprocessor, a connections expansion board and the desired sensors. Two micro DC motors are driven by a dual channel motor driver. The disposition of the components can be seen in Fig. 3

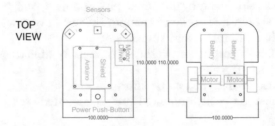

Fig. 3. CAD diagram with dimensions

3.3 Software

The software has in consideration that the robot is being developed for improving the students skills at some fields of robotics. Therefore, there are some blocks of code - namely Navigation, Control Loop and Controller - that must be developed by students and implemented in the main software of the robot.

So, NaSSIE has sensors, a processing unit and actuators that creates a cyclical loop chain through which the system operates. This allows the platform to sense its surroundings, avoid obstacles and navigate. Also, the platform has actuators, two DC micro motors, which enable the robot movement.

The architecture of the system is presented in Fig. 4. The green block has the sensoring part of the platform, the orange block corresponds to the actuation part of the platform and the purple block has the detailed processing unit part of the project. Inside this, the red blocks are the ones that students must develop on their own. It is noteworthy that the full system can be easily reconfigured to host new blocks, that can serve as an expansion of the system capabilities.

Analyzing the Figure, as inputs of the system, there are the sensors unit. As previously mentioned, the sensors will give information to the microprocessor that is used to understand the robot's surroundings. The speed controller has two modules: a higher level one, which decides which speeds to apply to each motor based on the values of linear and angular speed, defined on the state machine and a lower level one, that controls the speed of the motors using the

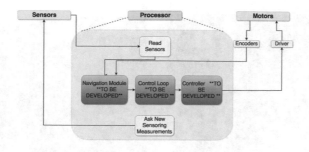

Fig. 4. System architecture

encoders readings. The controller block is also an important one, since one of the tasks that students have to do is to define which controllers to use for the vehicle, and to tune them up, considering the environment conditions and restraints. The output of this block is the reference speed to each motor controller. Finally, the processing unit asks for new measurements from the sensoring unit and the loop starts all over again.

4   Prototype

To serve as a use case, the robot presented in Fig. 5 was built. For the construction of the upper and lower bases with a very similar shape, with maximum size of approximately 110 × 100 mm. However, the lower one has two cuts to make room for the wheels. To decrease drift when the vehicle moves, wheels with a rigid center in plastic and an outer rubber tire were bought. Two 6 V micro DC motors with incorporated gearboxes were placed side by side. Initially, it was tested several configurations of gearboxes, however it was concluded that for purposes of speed testing, slippage and better response of the robot to impulses, it would be better to use a gearbox of 30:1. The commands for controlling these DC motors come from a dual motor driver. This driver receive commands from the main processing unit and control the motors speed. An Arduino Nano was selected, firstly because of its small size, which is important since the platform is intended to be small and easy to be acquired by anyone. The development community of this platform was also considered, since there are already numerous libraries that allow the use of the most varied components, using various types of communication. To assist in the distribution of the different analog and digital signals, as well as to facilitate the input of the encoder signals of the motors and sensor signals, an expansion shield for the Arduino Nano was used. The distance sensors use Time-of-Flight (ToF) technology and consist in a breakout board that has an integrated circuit with a laser transmitter and receiver that allows to measure relatively short distances. These sensors are used to know how far away the obstacles are from the front of the vehicle. For the system to be more reliable, three sensors were placed so that the robot can see how close the obstacles are to a 200 mm semicircle in front of it (Fig. 6).

Fig. 5. Robot

Fig. 6. Semi-circle of sensing

The measured data is sent through I2C and this information is sent directly to the Arduino. Finally, two Lithium-ion Polymer Battery of 1600 mAh, with an output voltage of 7.2 Volts were used. The complete list of the components used in this prototype is presented in table 1.

Table 1. Components for the robot construction

Component	Brand	Model	Brand reference
Batteries	CELLEVIA	Lithium-ion Polymer Battery 3.7 V 1600 mAh	N/A
Development board	Arduino	Nano	A000005
Shield	DFROBOT	Shield I/O Nano	DFR0012
Power button	Adafruit	Push-button Power Switch	1400
Distance sensors	POLOLU	VL6180X Time-of-Flight Distance Sensor	2489
Voltage regulator	POLOLU	5 V, 600 mA Step-Down Voltage Regulator D24V6F5	2107
Motor driver	POLOLU	TB6612FNG Dual Motor Driver	713
Motor	DFROBOT	Micro DC Geared Motor w/Encoder - 6V 530RPM 30:1	FIT0481
Wheels	DFROBOT	Wheel 42 × 19 mm	FIT0085

5   Approach and Results

In both subsections below, there are presented use cases where the robot was used by a group of students to compete on international competitions.

Robotic Day 2017

For this competition, the team has implemented a solution that use the three ToF sensors available on the platform to follow the inner walls of the circuit. The robot starts from the bottom right corner on a randomly attributed pit. Before the start, the pits are closed and the vehicle detects an obstacle using these sensors. When the pit is opened, the vehicle path is unobstructed and therefore, the vehicle accelerates in a straight line. After a defined amount of time, the vehicle starts approaching the inner wall until it detects the wall. It then proceeds to follow the wall on the left of the robot maintaining a fixed distance off the wall. When the sensor on the left does not detect a wall, it means that the robot has passed the straight wall on the left and so makes a 180 degrees turn until it senses the new wall on the left again. When the robot senses an obstacle on the front of the vehicle, it makes a 90 degrees turn until it only senses the wall on the left. This approach also allows to perform overtakes since this is similar to following the racetrack: the vehicle senses an obstacle at the front and turns right; then, when it does not sense an obstacle at the left, the vehicle turns until it is on its left; after overtaking the other vehicle it turns left again, since it does not sense an obstacle at its left; the vehicle returns to following the wall of the circuit.

During this competition, the robot has faced some challenges, namely with the ToF Sensors as well as the motors. Regarding the ToF sensors, they do not correspond to the specified distance measurements. Initially, when the sensors were tested, the values of the measured distance were always correct. However, in the competition, the sensors only read distance values below 10 cm. This was due to the wall's colours (red and green), since the distance to a black or white wall was read correctly.

Also, the micro-geared motors had an open gearbox which was subjected to external debris. This has caused the gearbox to malfunction, which makes one motor rotate slower than the other. To cope with this, the authors have changed the motors to a spare ones. Despite all the adversities, the team has reached a third place, performing 5 laps in 77.5 s.

Micromouse Portuguese Contest

As off today, this robot was not used in this competition since this will only happen in May, 2018. However, based on the efforts that are being performed by this year's team, it is expected that the robot will complete the maze in less than 2 min.

6   Conclusions and Future Work

In this paper it was presented the development of a new platform to introduce Navigation and Sensoring skills to new users. Therefore, this platform was designed using a simple approach to facilitate the use of the robot and to illustrate real problems in the field of Mobile Robotics, using a hands-on approach.

The developed robot was designed to be small and simple, using components that can be easily acquired and replaced. The robot construction also takes into account that its target audience is inexperienced and, therefore, has a robust design to prevent significant damages. This configuration serves the main purpose of this platform, that will be used in different educational applications.

This platform was tested in the Robotic Day 2017 where it fulfilled all the expectations providing a robot that was able to compete with other vehicles.

This experience allowed the participating students to acquire skills in the field of Mobile Robotics in a real use case with a platform suitable for this application, due to its simple design with independent blocks that facilitates parts replacement.

The developed platform can be used in other scenarios, namely different competitions like the Micromouse competition. This robot can also be used to illustrate concepts in classes, as a platform to encourage the curiosity about the field of Robotics, to teach basic concepts related with this field and in demonstrations like "Semana Profissão Engenheiro" and "Universidade Júnior", which are the divulgation programs from Faculty of Engineering of University of Porto, for highschool students.

As future work, it is planned to implement over-the-air programming and debugging to facilitate the development in this platform. Moreover, it is planned to add new sensors as needed for different applications.

Aknowledgements. This work is financed by the ERDF European Regional Development Fund through the Operational Programme for Competitiveness and Internationalisation - COMPETE 2020 Programme within project POCI-01-0145-FEDER-006961 and by National Funds through the FCT Fundação para a Ciência e a Tecnologia (Portuguese Foundation for Science and Technology) as part of project UID/EEA/50014/2013.

References

1. IEEE Micromouse Competition Guide (1993). http://www.tic.ac.uk/micromouse/guide/index.asp. Accessed 26 Nov 2017
2. Micromouse Competition (2017). http://www.micromouse.utad.pt/?page_id=504&lang=en. Accessed 26 Nov 2017
3. Prague Robotic Day 2017 (2017). http://roboticday.org/. Accessed 16 Nov 2017
4. Almeida, F.L., Terra, B.M., Dias, P.A., Gonçalves, G.M.: Transport with automatic guided vehicles in the factory of the future. In: 2010 IEEE Conference on Emerging Technologies and Factory Automation (ETFA), pp. 1–4. IEEE (2010)
5. Khurshid, J., Bing-Rong, H.: Military robots-a glimpse from today and tomorrow. In: 2004 8th Control, Automation, Robotics and Vision Conference. ICARCV, vol. 1, pp. 771–777. IEEE (2004)
6. Mubin, O., Stevens, C.J., Shahid, S., Al Mahmud, A., Dong, J.-J.: A review of the applicability of robots in education. J. Technol. Educ. Learn. **1**(209–0015), 13 (2013)
7. Wallach, B.A., Koselka, H.A., Gollaher, D.L.: Autonomous vacuum cleaner, 9 August 2005. US Patent 6,925,679

Education with Robots Inspired in Biological Systems

Nuno M. Fonseca Ferreira[1,2,3(✉)], Fernando Moita[1,2,3,4],
Victor D. N. Santos[3,4], João Ferreira[3], João Cândido Santos[3],
Frederico Santos[3], and Marco Silva[3]

[1] Knowledge Research Group on Intelligent Engineering and Computing
for Advanced Innovation and Development (GECAD) of the ISEP, IPP,
Porto, Portugal
[2] Institute of Engineering of Coimbra, Polytechnic Institute of Coimbra,
Coimbra, Portugal
nunomig@isec.pt
[3] Instituto de Engenharia de Sistemas E Computadores,
Tecnologia E Ciência, Porto, Portugal
[4] INESC TEC - Institute for Systems and Computer Engineering,
Technology and Science, Porto, Portugal

Abstract. This paper presents one methodology for teaching engineering students, which relies on open platform requiring basic knowledge of robotics, like mechanics, control or energy management. Walking robots are well known for being able to walk over rough terrain and adapt to various environments. Hexapod robots are chosen because of their better stability and higher number of different gaits. However, having to hold the whole weight of the body and a large number of actuators makes all walking robots less energetically efficient than wheeled robots. This platform endows students with an intuitive learning for current technologies, development and testing of new algorithms in the area of mobile robotics and also in generating good team-building.

Keywords: Education · Project based learning and robotics
Mechanism designs · Kinematics of robots · Robot motion

1 Introduction

Traditional robotics courses deal with kinematics, dynamics and control of manipulators [1–3]. Progress in scientific research and developments on industrial applications lead to the appearance of educational programs on robotics, covering a wide range of aspects such as kinematics, artificial intelligence and mechanical design. It is fundamental to involve students in the construction and development of robotic platforms, and in project-oriented learning in order to forest creative and divergent thinking, and promote acquisition of self-learning and practical skills [4–9]. For this reason we allows the students to develop different robotics platforms. The creation of working groups encourages autonomous research as well as teamwork, resulting in greater involvement of students and greater dedication, thus facilitating learning. The use of a robotic teaching platform allows the students a fast understanding of the theoretical topics, and

W. Lepuschitz et al. (Eds.): RiE 2018, AISC 829, pp. 207–213, 2019.
https://doi.org/10.1007/978-3-319-97085-1_21

the use of Arduino-based open platform allows students to easily evolve acquired knowledge for project-based learning [4–7]. In this paper, we will focus on walking robots. The legged robots, compared with wheeled or tracked robots, have much superior movement characteristics due to greater ability to adapt to rough terrain [10]. Just like most of the animals use legs to move and adapt to nature, have good adaptability to various environments [11]. However, autonomous movement over rough terrain is a complex task. All walking robots require a high level of research, control, and parameter configuration [12] or transportation [13]. Thus, different legged locomotion control methods are needed to perform well the desired task of the system during walking. All walking robots tend to use much energy due to having to hold the whole weight of the body and all driving parts [14]. This is very important, especially for autonomous robots, because decreasing power consumption is one of the possibilities to increase work time of a robot without changing or charging the batteries [15]. In order to minimize hexapod robot's energy consumption when walking on even terrain, an energy consumption model must be established [16]. In order to reach demands for hexapod robot energy consumption, an algorithm for torque distribution with an energy consumption model [17]. Choosing different gaits and gait parameters are the most important and general factors to be taken into consideration. Thus, the analysis of turning gait parameters of a six-legged robot while doing turns was done [18]. Results showed that, unlike tripod gait which is the fastest, but also very unstable, wave gait is the most energetically efficient and the most stable choice. Another way to minimize power consumption is by observing animals in nature and applying same movement methods for robots. Most of the animals try to find a position in which they would use the least amount of energy, but still completing the needed task. Even humans have the same tendencies.

2   Mathematical Model

The chassis has 6 protruding parts through which has all the limbs are attached. The material its light weight and strength. Each leg has three servo motors to allow motion in three dimensions. Before any implementation we test in virtual environment the model of the proposed hexapod with similar characteristics are mention in [16–20] (Fig. 1).

Fig. 1. Spider with proposed circular shape hexapod mechanical structure

2.1   Forward Kinematics

Homogenous transformation matrix for a given leg can be determined by using the Denavit-Hartenberg (H-D) Convention [21]. This representation is composed of four transformations.

Table 1. Link parameters

$link_i$	α_i	a_i	$\cos \alpha_i$	$\sin \alpha_i$	d_i	Variable
1	$\pi/2$	0	0	1	0	θ_1
2	0	L_1	1	0	0	θ_2
3	0	L_2	1	0	0	θ_3

$$H = \begin{bmatrix} R & d \\ 0 & 1 \end{bmatrix} \qquad (1)$$

where R = Rotational matrix, d = Translational matrix, Fig. 2 shows the leg of the hexapod with three revolute joints for creating 3 RRR manipulator with 3 DOF and the joints are the representatives of θ_1, θ_2 and θ_3. The general form of H-D convention is mentioned as:

$$A_i = \begin{bmatrix} \cos \theta_i & -\sin \theta_i \cos \alpha_i & \sin \theta_i \cos \alpha_i & a_i \cos \theta_i \\ \sin \theta_i & \cos \theta_i \cos \alpha_i & -\cos \theta_i \cos \alpha_i & a_i \sin \theta_i \\ 0 & \sin \alpha_i & \cos \alpha_i & d_i \\ 0 & 0 & 0 & 1 \end{bmatrix} \qquad (2)$$

By using Table 1, the matrix H-D for the joints is given in (4), (5), (6):

$$T_3^0 = A_1^0 A_2^1 A_3^2 \qquad (3)$$

Transformation metrics for each joint can be define as following by using (2).

$$A_1^0 = \begin{bmatrix} \cos \theta_1 & 0 & \sin \theta_1 & 0 \\ \sin \theta_1 & 0 & -\cos \theta_1 & 0 \\ 0 & 1 & 0 & 0 \\ 0 & 0 & 0 & 1 \end{bmatrix} \qquad (4)$$

$$A_2^1 = \begin{bmatrix} \cos \theta_2 & -\sin \theta_2 & 0 & l_2 \cos \theta_2 \\ \sin \theta_2 & \cos \theta_2 & 0 & -l_2 \sin \theta_2 \\ 0 & 0 & 1 & 0 \\ 0 & 0 & 0 & 1 \end{bmatrix} \qquad (5)$$

$$A_3^2 = \begin{bmatrix} \cos \theta_3 & -\sin \theta_3 & 0 & l_3 \cos \theta_3 \\ \sin \theta_3 & \cos \theta_3 & 0 & -l_3 \sin 3 \\ 0 & 0 & 1 & 0 \\ 0 & 0 & 0 & 1 \end{bmatrix} \qquad (6)$$

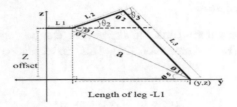

Fig. 2. Model for inverse kinematics in z y coordinates

Parametric equations are deduced by making use of (3).

$$x = L_2 \cos \theta_1 \cos(\theta_2 + \theta_3) + L_3 \cos \theta_1 \cos \theta_2$$
$$y = L_2 \sin \theta_1 \cos(\theta_2 + \theta_3) + L_3 \sin \theta_1 \cos \theta_2 \qquad (7)$$
$$z = L_2 \sin(\theta_2 + \theta_3) + L_3 \sin \theta_2$$

2.2 Inverse Kinematics

There are different methods to derive inverse kinematics equations, the geometric method has been used and the representation of 3-D design on 2-D, design into two figures. One provides the top view of leg motion in x, y coordinates and the other the Z axis. The θ_1 can obtain by inverse tangent function as shown in Fig. 3 this approach can be verify from [10, 11] and book reference [14].

$$\theta_1 = \tan^{-1}\left(\frac{y}{x}\right) \qquad (8)$$

The length of the leg is the hypotenuse of the triangle in Fig. 3, and it can be calculated by using Pythagorean Theorem.

$$Length\ of\ Leg = \sqrt{x^2 + y^2} \qquad (9)$$

Figure 3 represents the side view of a leg in y, z coordinates and θ_2 and θ_3, the resultant vector 'a' divides the total angle between the lines L_3 and L_1 + Length of leg into two angle and its magnitude can be given by:

$$a = \sqrt{(Length\ of\ Leg - L_1)^2 + Z^2} \qquad (10)$$

φ_1 is the angle between L_2 and resultant vector 'a', and φ_4 is present among the perpendicular line on y axis and resultant vector 'a', than we can determine θ_2 as:

$$\theta_2 = \varphi_1 + \varphi_4 - \frac{\pi}{2} \qquad (11)$$

Using the law of cosines we ca write θ_2:

$$\theta_2 = \cos^{-1}\left(\frac{a^2 + l_2^2 + l_3^2}{2aL_2}\right) + \tan^{-1}\left(\frac{Length\ of\ Leg - L_1}{Z\ offset}\right) - \frac{\pi}{2} \qquad (12)$$

θ_3 can also be calculated from Fig. 3:

$$\theta_3 = \pi - \cos^{-1}\left(\frac{-a^2 + l_2^2 + l_3^2}{2L_3L_2}\right) \qquad (13)$$

3 Control of a Walking Robot

Motion control should provide leg movements, in the control we must consider, the control gait (the sequencing of leg movement), the control of foot placement, and the control body movement for the supporting legs. For the legs we have two major states, the stance (in which the leg is in the ground), and the other (when it is in the air) moving to a new position. The fly phase has three main components, the lift phase (when the leg is leaving the ground), the transfer between positions (when the leg is moving to a new position), and the last one (the landing of the leg), the smooth placement on the ground. The gaits determine the sequence of configurations of the legs, gaits can be divided into two main classes: periodic gaits, which repeat the same sequence of movements, and non-periodic or free gaits, which have no periodicity in the control, and could be controlled by layout of environment. There are several different methods of moving the six legs of a walker robot to create locomotion. These types include ripple, wave, and tripod gait. The wave gait is the most stable gait and allows smooth forward leg movements, when all legs on one side move forward at same time, motion starts at the rear leg and progresses forward. The tripod gait is both statically and dynamically stable where three legs are always on the ground.

4 Hexapod Robot

The hexapod robot used in this experiment is shown in Fig. 3, the sonar is used to detect obstacles. Main parameters of the robot are: weight M ≈ 1.5 kg; body width Lw = 83 mm; body length L = 193 mm; each leg L_{coxa} = 68 mm, L_{femur} = 80 mm, L_{tibia} = 105 mm.

Fig. 3. Hexapod robot model used in the experiment.

5 Walking vs Running

Motion of a legged system is called walking if in all instances at least one leg is supporting the body. If there are instances where no legs are on the ground it is called running, Walking can be statically or dynamically stable, running is always dynamically stable. Stability means the capability to maintain the body posture given the control patterns, statically stable walking implies that the posture can be achieved even if the motion is stopped at any time, without loss of stability, dynamic stability implies that stability can only be achieved through active control of the leg motion. These experiments were carried out in a real environment, where the students tested the robot walking vs running, and were allowed to understand the concepts of stability.

6 Research Work

Some research work [10, 11], observe the static characteristics of a hexapod walking robot. Energy consumption dependence on number of legs on surface. Three different cases were studied: six, three and zero legs set on surface. Results showed that energy consumption increases as the number of legs supporting the weight decreases. The Energy consumption stability depends on how fast legs are transferred during the motion.

7 Conclusions and Future Work

Summarizing, the proposed robotic platform allows the students, start to improve and develop new ideas, also this type of robotic platform it is very good for serious research. The teaching approach with Educational robotics promotes the future success of our students, we can say that when we observed the students during the class, they were fully engaged and were having fun. Based on their feedback, we were surprised to note that they built two identical robotic platforms, also this described the activity of the students when they are involved with robots.

References

1. Ferreira, N.M.F., Machado, J.A.T.: RobLib: an educational program for robotics. In: Proceedings of 6th International Symposium on Robot Control, Vienna, Austria, pp. 563–568 (2000)
2. Ferreira, N.M.F., Machado, J.A.T.: Teaching robot modelling and control with RobLib. In: Proceedings of Controlo 2000, 4th Portuguese Conference on Automatic Control, 4–6 October, Guimaraes, Portugal, pp. 406–411 (2000)
3. Verner, I.M., Waks, S., Kolberg, E.: Educational robotics: an insight into systems engineering. Eur. J. Eng. Educ. 24(2), 201–212 (1999). https://doi.org/10.1080/0304379990 8923555

4. Couceiro, M.S., Figueiredo, C.M., Luz, J.M.A., Ferreira, N.M.F., Rocha, R.P.: A low-cost educational platform for swarm robotics. Int. J. Robots Educ. Art, IJREA 2(1), 1–15 (2012). https://doi.org/10.4156/ijrea.vol2.issue1.1
5. Couceiro, M.S., Ferreira, N.M.F., Rocha, R.: Multi-robot exploration based on swarm optimization algorithms. In: ENOC 2011 - 7th European Nonlinear Dynamics Conference, 24–29 July, Rome, Italy (2011)
6. Sousa, I.M., Barbosa, A.R., Couceiro, M.S., Figueiredo, C.M., Ferreira, N.M.F.: Exploiting the development of robotic hands - a survey and comparison on different technologies. In: 2nd International Conference on Serious Games and Applications for Health - SeGAH 2013, 2–3 May, Vilamoura, Portugal (2013)
7. Couceiro, M.S., Ferreira, N.M.F., Machado, J.A.T.: Hybrid adaptive control of a dragonfly model. J. Commun. Nonlinear Sci. Numer. Simul. 17(2), 893–903 (2012). Elsevier
8. Vital, J.P.M., Rodrigues, N.N.M., Couceiro, M.S., Figueiredo, C.M., Ferreira, N.M.F.: Fostering the NAO platform as an elderly care robot - first steps toward a low-cost off-the-shelf solution. In: 2nd International Conference on Serious Games and Applications for Health - SeGAH 2013, 2–3 May, Vilamoura, Portugal (2013)
9. Couceiro, M.S., Luz, J.M.A., Figueiredo, C.M., Ferreira, N.M.F.: Modeling and control of biologically inspired flying robots. Robotica, vol. 30, no. 1, pp. 107–121. Cambridge University Press (2012)
10. McGhee, R.B., Iswandhi, G.I.: Adaptive locomotion of a multilegged robot over rough terrain. IEEE Trans. Syst. Man Cybern. SMC-9(4), 176–182 (1979)
11. Kar, D.C., Issac, K.K., Jayarajan, K.: Gaits and energetics in terrestrial legged locomotion. Mech. Mach. Theory 38, 355–366 (2003)
12. Jin, B., Chen, C., Li, W.: Power consumption optimization for a hexapod walking robot. J. Intell. Rob. Syst. 71, 195–209 (2013)
13. Buchli, J., Kalakrishnan, M., Mistry, M., Pastor, P., Schaal, S.: Compliant quadruped locomotion over rough terrain. In: IEEE/RSJ International Conference on Intelligent Robots and Systems, pp. 814–820 (2009)
14. Shekhar, S.R., Pratihar, D.K.: Effects of turning gait parameters on energy consumption and stability of a six-legged walking robot. Robot. Auton. Syst. 60, 72–78 (2011)
15. de Santos, P.G., Garcia, E., Ponticelli, R., Armada, M.: Minimizing energy consumption in hexapod robots. Adv. Robot. 23, 681–704 (2009)
16. Roy, S.S., Pratihar, D.K.: Effects of turning gait parameters on energy consumption and stability of a six-legged walking robot. Robot. Auton. Syst. 60, 72–82 (2012)
17. Luneckas, M., Luneckas, T., Udris, D.: Hexapod walking robot energy consumption dependence on the number of legs set on the surface. In: Proceedings of the 8th International Conference on Electrical and Control Technologies ECT – 2013, pp. 25–28 (2013)
18. Luneckas, M., Luneckas, T., Udris, D., Ferreira, N.M.F.: Hexapod robot energy consumption dependence on body elevation and step height. J. Elektron. Elektrotech. 20, 7–10 (2014)
19. Brockmann, W., Maehle, E., Mösch, F.: Organic fault tolerant control architecture for robotic applications. In: 4th IARP/IEEE-RAS/EURON Workshop on Dependable Robots in Human Environments, Nagoya, Japan (2005)
20. Ollervides, J., Orrante-Sakanassi, J., Santibanez, V., Dzul, A.: Navigation control system of walking hexapod robot. In: 2012 IEEE Ninth Electronics, Robotics and Automotive Mechanics Conference (CERMA), November 2012
21. Spong, M.W., Hutchinson, S., Vidyasagar, M.: Robot Modeling and Control. Wiley, Hoboken (2005)

Programming Environments
and Languages

Tailoring a ROS Educational Programming Language Architecture

Karen Tatarian[1], Samuel Pereira[2], Micael S. Couceiro[2,3]([✉]),
and David Portugal[2]

[1] American University of Beirut, Beirut 1107 2020, Lebanon
tatariankaren@gmail.com
[2] Ingeniarius, Lda., Rua Coronel Veiga Simão, Edifício CTCV,
3025-307 Coimbra, Portugal
{samuel,micael,davidbsp}@ingeniarius.pt
[3] Institute of Systems and Robotics, University of Coimbra - Pólo II,
Rua Sílvio Lima, 3030-290 Coimbra, Portugal
micaelcouceiro@isr.uc.pt
https://www.linkedin.com/in/karen-tatarian-52a790128/
http://www.ingeniarius.pt/, http://www.isr.uc.pt/~micaelcouceiro

Abstract. With its impressive rise in popularity within education at all levels, Robotics is a rapidly growing field merging science, technology, engineering and mathematics (STEM). Nevertheless, the lack of standards in educational robotics has led to several issues, namely little (or almost no) code reuse between educational robotics curriculum from different schools, large dependency on proprietary solutions, and endless paradigm shifting between text, visual and flow programming languages. This paper proposes a novel educational programming language architecture to teach students how to program robots. The architecture combines the Robot Operating System (*ROS*) quasi-standard with the *Snap!* visual programming language, targeting students from primary education to high school. As opposed to the limited alternatives available in ROS, the solution proposed does not require the acquisition of any robotic platform, running directly on the browser, and benefiting from the power of the internet to program ROS-enabled real and simulated robots.

Keywords: Educational robotics · Programming · ROS architecture

1 Introduction

Robotics is a growing field that merges science, technology, engineering and mathematics (STEM) [1]. This unique multidisciplinary nature, and the science fiction revolving around it, has astonishingly increased its popularity within education at all levels [2]. Educational robotics has become one of the most popular activities to increase and facilitate the engagement of children in the teaching-learning process of STEM-related topics.

© Springer Nature Switzerland AG 2019
W. Lepuschitz et al. (Eds.): RiE 2018, AISC 829, pp. 217–229, 2019.
https://doi.org/10.1007/978-3-319-97085-1_22

Despite increasing instructional programs, platforms, educational resources and pedagogical philosophy of educational robotics, the lack of standards, which resulted in the inevitable adoption of several proprietary software tools and robots, such as *LEGO Mindstorms NXT* or *Microsoft Robotics Studio* [3], led to some issues typically shared among general robotics as well, including [4]:

- **code re-usability:** there is little (to almost none) code reuse between educational robotics curriculum from different schools, slowing down knowledge and skills sharing among the teacher and student communities;
- **proprietary solutions:** the large dependency on proprietary software and hardware (*e.g.*, interdependency between programming environment and robotic platforms), requires schools to affiliate with companies under proprietary license agreements, typically with overpriced solutions;
- **programming paradigms:** an endless paradigm shift between text, visual and flow programming languages, requires students to "rewire their brains" in order to grasp the different languages and adequately use them for problem solving.

Several frameworks were proposed to tackle the lack of code re-usability, the need for proprietary solutions and the shifts from programming paradigms, from which the Robot Operating System (*ROS*) stood out as a *de facto* standard [5]. ROS[1] is an open-source meta-operating system for robots, providing hardware abstraction, low-level device control, message-passing between processes, and package management. ROS encompasses multiple tools and libraries, constantly updated by the robotic community, to help software developers create robot applications. The number of ROS-enabled robots has been increasing at a fast pace, accounting for hundreds of educational and professional solutions[2].

Naturally, one would expect that by employing ROS in the early stages of education, students would get in touch with the most complete and popular robotic framework, thus being engaged in state-of-the-art STEM-related topics. Furthermore, as almost all universities with robotics research and many high-end robotics companies have been adopting ROS, this would endow students with the possibility to handle a wide variety of robots in the future and grow into adults with STEM careers to foster a healthy economy [6].

Nonetheless, mastering ROS has proved to be non-trivial at pre-university level, as it requires *a priori* programming and Linux skills. In several cases, the documentation of the default wiki pages lack support, and most ROS-users rely on books to start building their first robotic routines [7]. In addition, there are many ROS-compatible simulators, such as *Gazebo*, *MORSE* and *Stage* [8]. However, to get started with these simulators is not an easy task. Unlike commercial simulators with inbuilt functionalities and a rich graphical user interface (GUI) to prototype and program the robots, such as *V-REP* and *Webots*, simulations involving ROS are done mostly through coding.

Many visual programming tools with educational purposes were created for programming, such as *Scratch*, *Snap!*, *Alice*, and *Blockly*. Provided free of

---

[1] http://www.ros.org.

[2] http://robots.ros.org/all.

charge, Scratch was a project, developed by MIT Media Lab, aiming at helping young children think creatively, reason methodically, and work collabora-
tively[3]. Scratch can communicate with ROS using a Python interface[4]. Being
the extended reimplementation of Scratch, Snap! is a visual, drag-and-drop programming language[5]. Similarly, Alice is a block-based programming tool to learn
object-oriented programming, developed by Carnegie Mellon University[6]. One
of the most recent programming tools published in 2015, *robot_blockly* was developed to make the process of prototyping robot systems simpler by allowing the
control of a robot or drone using ROS[7]. Likewise, *Robokol* was presented as a
graphical environment based on both ROS and Snap!, focusing on the description
of applications, namely robots for therapists of autism [16].

However, the motivation for the ongoing research are the visible drawbacks
that each of the above mentioned tools, as well as others, present. Firstly, regarding the utilization of robotics middleware, Scratch, Snap!, and Alice do not allow,
as is, the user to operate a robot; they are simply used for educational programming purposes. Only robot_blockly and Robokol use robotics middleware. Yet,
blocks of robot_blockly can only be used on robots created by *Erle Robotics,
Inc.*, while the authors of Robokol focus in the applicability of the solution to
specific case studies, without delineating the ROS-Snap! architecture. Thirdly,
despite the ease in building initial development environments, Scratch, Snap!,
and the Scratch interface for ROS do not allow the demonstration of the source
code even though the operation specified by the block is highlighted during execution. On the other hand, Alice can give the user the option to see the Java
source code and Blockly can be integrated with programming languages, such
as Python and JavaScript [17].

Moreover, many of the tools currently found are block-based coding tools,
which clearly raises the question on their limitations. Published in 2015, a study
on the effectiveness of block-based programming on high school students was
done by Weintrop and Welinsky [18]. The study concluded that even though
the advantages provided by block-based programming, such as natural language
labels on the blocks, shapes and colors, and the drag-and-drop mechanisms,
made the students generally find it easier, they were still able to point out many
drawbacks. These limitations include a lack in expressive power, referring to the
set of possibilities for programming with such tools in comparison to text-based
alternatives setting limits for executing complex functions. In addition, they were
faced with challenges in authoring larger and more sophisticated programs since
it took time, and an extensive number of blocks, to compose a program using
a block-based interface in comparison to the text-based alternatives. Finally,
one of the most significant limitations was the lack of authenticity, since these

---

[3] https://scratch.mit.edu.

[4] http://wiki.ros.org/scratch.

[5] http://snap.berkeley.edu.

[6] http://www.alice.org/index.php.

[7] http://wiki.ros.org/blockly.

programming tools are remotely distant from the conventional non-educational programming context.

More recently, in the Robobo project[8], a smartphone-based robotic platform was proposed, encompassing a programming environment and incorporating lessons [15]. The hardware comprises of a wheeled base, containing high-level sensors and control for base actuators, as well as a smartphone. The sensors, communication, and actuation heavily depend on the smartphone hardware. Additionally, the smartphone is provided by the students themselves, which may pose an issue, as students might not have smartphones, specially in under-developed countries, or are not allowed to use them in classrooms. The software also requires an Android smartphone, which limits its use. Regarding the Robobo software, despite the four different programming paradigms offered (Java, Scratch, Block, and ROS), it is centered in a Robobo module and framework for libraries; which raises authenticity issues, as mentioned by Weintrop and Wilensky [18]. Finally, the lessons are limited to ScratchX blocks, limiting the students to a specific software.

1.1 Statement of Contribution

This paper proposes a novel educational programming language architecture to teach students how to program robots. Similarly to the Robobo project , the architecture proposed combines the ROS quasi-standard with the Snap! visual programming language, targeting students from primary education to high school (up to the level 3 of the International Standard Classification of Education, ISCED). Nevertheless, instead of restraining students to a specific type of robot and local development, the proposed architecture allows students to develop their robotic routines over the internet, using *Snap!* block-based programming directly on the browser, benefiting from the power of the internet to program real and simulated ROS-enabled robots.

The next section summarizes the past educational activities that our research group was associated with and the take home messages yielded in order to establish the guidelines defining the proposed architecture. Section 3 presents the ROS educational programming language architecture, describing its most relevant components, namely the ROS-enabled robot used to evaluate it, `rosbridge`, `roslibjs` and *Snap!*. A preliminary evaluation of the functional characteristics of the architecture is carried out in Sect. 4. Finally, Sect. 5 ends with some final remarks on the adopted approach and future directions.

2 Lessons Taken from Previous Initiatives

An increasing number of teachers are integrating robotics in their classrooms in order to further engage students into the educational process, thus resulting in better learning outcomes. Although educational robotics does not exist as a

---

[8] http://www.theroboboproject.com.

dedicated subject anywhere in the world, teachers from all levels of education have found many and different ways to integrate robotics into their lessons.

The authors of this paper have been mostly using robotics to teach related engineering disciplines, in public and private domains, and aimed at high and higher education. Besides the typical higher education modules, such as digital systems, programming and artificial intelligence, the authors were involved in the organization of ISEC Summer School from 2010 to 2012, at the Engineering Institute of Coimbra (ISEC). The summer course brought together between 30 to 45 high-school students to each of the three versions of the course, instructing them how to build and program a simple differential robot during a two-week intensive programme of 6 hours/day. The students attended multiple modules, including digital systems, electronics, mechatronics, and microprocessors (namely Arduino). In 2016, the experience and outcome provided by ISEC Summer Course led to the creation of an international programme, named Robotics Craftsmanship International Academy (RobotCraft[9]). The authors were again involved in the organization of this event, now represented by the Ingeniarius, Ltd. company, in close collaboration with the University of Coimbra. Unlike before, this initiative was seen as a collective internship, consisting of higher school students coming from all over the world. The first edition, in 2016, received 65 students from 18 different countries, while the second edition, held in the summer of 2017, received 85 students coming from 21 countries. Moreover, RobotCraft ran for two consecutive months, encompassing similar engineering disciplines as in ISEC Summer School, such as electronics, mechatronics and microprocessors, as well as some others, such as computer-aided design (CAD), the Robot Operating System (ROS) and artificial intelligence.

Even though these initiatives targeted an older population than the one aimed in this paper, they shared a similar principle: to increase both engagement and skills in STEM-related topics. The programmes addressed science by teaching artificial intelligence, technology through the integration of sensing, actuation and communication devices, engineering by exploring the mechatronics design of robots, and mathematics by developing the kinematic model and control architecture. Programming acted as the "glue" connecting these STEM fields and, regardless of the level of the students, being from high school or university, most struggled with coding, namely to understand the language syntax, libraries, programming paradigms, etc. Due to its importance and the challenge it represents, this paper presents a ROS educational programming language architecture, designed to introduce robot programming in the early stages of education, while still benefiting from the power of ROS.

3   ROS Educational Programming Language Architecture

The proposed architecture is schematically represented in Fig. 1. The key feature of the architecture is to provide a bridge between an educationally-oriented programming language, in this case *Snap!*, with the ROS framework. The bridge

---
[9] http://robotcraft.ingeniarius.pt.

enables both local development using the wireless local area network (WLAN), and over the internet, through a virtual private network (VPN). It is noteworthy that, even though we have adopted a specific robot in this work, the architecture is robot-agnostic, and other platforms can be easily integrated in the architecture. The following sections describe every major component of the proposed ROS educational programming language architecture.

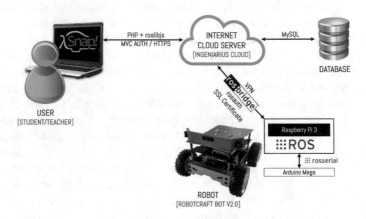

Fig. 1. Overview of the proposed ROS educational programming language architecture.

3.1 RobotCraft Bot V2.0: ROS-enabled Robot

Educational robots come in many forms and the market currently provides several solutions [9]. From simplistic teleoperated manipulators, such as the *Robotic Arm Edge*, to humanoid and more expensive solutions, such as *SoftBank Robotics' NAO*, schools have access to several off-the-shelf robots. Nevertheless, not every solution meets all the necessary requirements to be used for educational purposes, namely affordability, energy autonomy, communication interface, sensory system and processing power [10]. Despite being considered as the least relevant requirement of an educational robot in previous works, *e.g.* [9,10], processing power should not be underestimated. In order to run ROS, and to avoid integrating personal computers as the main source of robot's decision-making (*e.g.*, [9,11,12]), which would lead to an incremental cost, educational robots should be equipped with a ROS-enabled processor. Propitiously, robotic developers seized the opportunities offered by embedded system providers over the past few years, such as *Raspberry Pi*, *BeagleBone*, *Dragonboard*, *Odroid*, *Pine64*, *Udoo* and *Nvidia Jetson* boards.

The educational programming language architecture proposed was evaluated with *RobotCraft Bot v2.0*; an educational mobile robot developed within the context of RobotCraft 2017. As seen before, since we are following a ROS-based

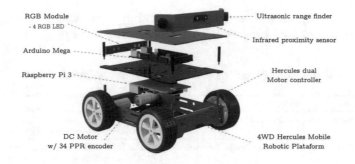

RGB Module
- 4 RGB LED

Arduino Mega

Raspberry Pi 3

Ultrasonic range finder

Infrared proximity sensor

Hercules dual
Motor controller

DC Motor
w/ 34 PPR encoder

4WD Hercules Mobile
Robotic Plataform

Fig. 2. The *RobotCraft Bot v2.0* robot structure.

approach, this is a hardware-independent architecture and can, therefore, be used in any ROS-enabled mobile robot.

RobotCraft Bot v2.0 follows a software and hardware architectures similar to other robots previously developed by the team and other ROS integrators, such as *TraxBot*[10] and *FORTE-RC*[11]. *RobotCraft Bot v2.0* is equipped with an *Arduino Mega* board, with access the motor encoders and other information from the power motor driver, the *Hercules Dual Motor Controller*, being also able to send commands to the motors, read sonar and infrared information, and exchange messages with a ROS-enabled *Raspberry Pi 3* (*cf.* Fig. 2). The low-level processing module (*Arduino*) communicates with the high-level (*Raspberry Pi 3*) using the `rosserial_arduino`[12] package. `rosserial_arduino` provides a ROS communication protocol that works over serial communication, thus allowing the low-level processing of the robot to be a full fledged ROS node which can directly publish and subscribe to ROS messages.

Table 1 depicts the list of topics published and subscribed between the high-level processing (*i.e.*, ROS side) and the low-level processing (*i.e.*, the *RobotCraft Bot v2.0 Arduino* driver). Generally, topics are named buses over which nodes (routines) exchange messages, being typically used for continuous data streams, as sensor data, robot state, among others. More specifically for the presented setup, messages can be seen as variables one intends to exchange between ROS and the robot, allowing user routines, or nodes, that are interested in robot data, to subscribe to the relevant topic where the messages are being exchanged (*e.g.*, `/distance_sensors`); contrariwise, nodes that generate data can publish to the relevant topic (*e.g.*, `/cmd_vel`).

As can be seen, the ROS paradigm of publishing and subscribing topics involving multiple nodes, is rather complex for non-experienced programmers and non-Linux users. For that reason, introducing ROS, and inevitably robotic programming concepts, for students up to ISCED level 3 can be an intimidat-

---

[10] http://wiki.ros.org/mrl_robots.

[11] http://wiki.ros.org/ingeniarius/ForteRC.

[12] http://wiki.ros.org/rosserial_arduino.

ing task, thus requiring a sequence of procedures that starts by employing a middleware abstraction layer to ROS, denoted as `rosbridge`.

Table 1. ROS topics provided by the *RobotCraft Bot v2.0*.

Msg. name	Msg. type	Description
`/distance_sensors`	std_msgs::Float32MultiArray	Subscribe ultrasonic left sensor, IR front sensor, and ultrasonic right sensor (m)
`/velocities`	std_msgs::Float32MultiArray	Subscribe linear (m/s) and angular (rad/s) speed, and left and right wheel speeds (m/s)
`/odom`	geometry_msgs::Pose2D	Subscribe the position (meters) and orientation (rad)
`/RGB_leds`	std_msgs::UInt8MultiArray	Publish the color of an array of four RGB LEDs
`/cmd_vel`	geometry_msgs::Twist	Publish the linear (meters/second) and angular (rad/s) velocities
`/initial_pose`	geometry_msgs::Pose2D	Publish the initial position (m) and orientation (rad)

3.2   rosbridge JSON API to ROS

The emergence of a wide range of robotic platforms and architectures have been calling upon new developments in robot middleware that might consider interoperability and code reuse; prerequisites that are necessary to the large scale development and deployment of robotic solutions. However, most robot middleware architectures are built for roboticistis, well-versed in the low-level systems programming and complex control and decision algorithms [13]. `rosbridge`[13] is a *JavaScript Object Notation* (JSON) *Application Programming Interface* (API) for ROS that allows the integration of ROS-enabled robots into non-ROS systems, with the intent to foster robot availability and accessibility.

`rosbridge` provides an additional level of abstraction on top of ROS, treating it as a "back end". This scales down the need to master the ROS programming paradigm. The `rosbridge` protocol is essentially a specification for sending JSON based commands to ROS, hiding its complexity using a simple socket serialization protocol, on top of which developers of all levels of experience can create applications. Put it differently, `rosbridge` allows simple message handling over both HTML5 websockets and standard POSIX IP sockets, thus allowing to exploit this paradigm in any language that supports IP sockets, which nowadays would mean all of them.

Additionally, one can leverage from *Robot Web Tools*[14], which already provides a javascript library that allows interfacing with ROS from the browser – `roslibjs`.

---

[13] http://wiki.ros.org/rosbridge_suite/.
[14] http://robotwebtools.org.

3.3   roslibjs Javascript Library

The `roslibjs` client library was developed to address a wide range of web-based ROS use cases and robotic applications [14]. Its design enables easy and flexible development of ROS-enabled robot routines, avoiding both a large monolithic structure and a large package system. `roslibjs` includes routines for common ROS and robotics tasks, such as publishing and subscribing topics, transform (TF) clients, Unified Robot Description Format (URDF) parsers, and common matrix/vector operations.

For instance, using `roslibjs`, a simple javascript client to publish velocity commands through the traditional ROS topic `/cmd_vel`, can be written as:

1. Connect to ROS:
```
var ros = new ROSLIB.Ros({
   url : 'ws://192.168.1.1:9090'
});
```

2. Create topic:
```
var cmdVel = new ROSLIB.Topic({
   ros : ros,
   name : '/cmd_vel',
   messageType : 'geometry_msgs/Twist'
});
```

3. Publish topic:
```
var twist = new ROSLIB.Message({
   linear : {
      x : 0.3,
      y : 0.0,
      z : 0.0
   },
   angular : {
      x : 0.0,
      y : 0.0,
      z : -0.2
   }
});
cmdVel.publish(twist);
```

Even by employing `rosbridge` and `roslibjs` on top of ROS-enabled robots, developers need to understand the build and transportation mechanisms, ROS concepts (*e.g.*, topics, nodes, services, etc.), and they still need to implement their own javascript projects in order to avoid using predefined simplistic routines, such as teleoperation or sensor data streams. Considering the challenge that this represents for the target population, with little to no "coding skills", an easy-to-use front-end should be adopted, allowing children to manipulate robotic program elements graphically, rather than by specifying them textually. Among the multiple existing choices as an educational visual programming language, the block-based programming tool *Snap!* offers the necessary flexible and intuitive development, while fostering code re-usability.

3.4   Snap! Blocks-Based Programming

Snap! is an HTML5 reimplementation of Scratch. It is a free, blocks-based educational graphical programming language that runs directly in the browser, thus supporting any OS. It extends Scratch by adding higher-level computer science concepts (*e.g.*, data structures) and providing the possibility to create custom blocks. Snap! was built to create interactive animations, games and stories using a simple drag-and-drop visual interface that can be mastered by virtually any 8 to 16 years old student.

In this work, we extend Snap!'s usability as a ROS educational programming language, employing mathematical and computational ideas to the programming of ROS-enabled platforms. For that purpose, several blocks were created, which call upon `roslibjs` functions, bridging the simplistic block-based programming offered by Snap! with the complex text-based programming inherent to ROS. Table 2 overviews the blocks created, presenting a brief description and the `roslibjs` functions called by each one of them.

Table 2. Examples of Snap! blocks created for interfacing with ROS.

Snap! block	Type	Description
ROS init	Control	Connects to the ROS-enabled robot
ROS move forward [x]	Motion	Sends a positive linear.x velocity to the robot
ROS move backwards [x]	Motion	Sends a negative linear.x velocity to the robot
ROS rotate left [z]	Motion	Sends a positive angular.z velocity to the robot
ROS rotate right [z]	Motion	Sends a negative angular.z velocity to the robot
ROS move stop	Motion	Sends a null linear.x and angular.z to the robot
ROS wheel left	Sensing	Receives the real left wheel velocity
ROS wheel right	Sensing	Receives the real right wheel velocity
ROS angular speed	Sensing	Receives the real angular.z speed reported by the wheels
ROS linear speed	Sensing	Receives the real linear.x speed reported by the wheels
ROS front sensor	Sensing	Receives the range reported by the front IR sensor
ROS left sensor	Sensing	Receives the range reported by the left ultrasonic sensor
ROS right sensor	Sensing	Receives the range reported by the right ultrasonic sensor

4   A Simple Scenario: Teleoperation

This section presents a simple application to teleoperate the *RobotCraft Bot v2.0* platform directly over the Snap! programming interface. Any other robot could be used as long as it is integrated in ROS and subscribes the `/cmd_vel` topic. Figure 3 presents the block-based code that allows to control the platform directly over the browser, using the WASD keys typically used in video games (W to move straight, A to turn left, S to move backwards, and D to turn right)[15].

---

[15] A demonstrative video can be seen at: https://www.youtube.com/watch?v=T9cAix LaW-8.

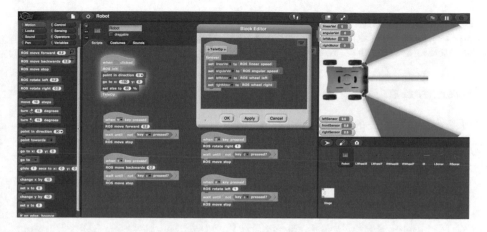

Fig. 3. Snap! GUI for teleoperation of the *RobotCraft Bot v2.0* platform over ROS.

Besides controlling the platform, the interface allows to observe the state of the ultrasound and infrared sensors by subscribing to the topic `/distance _sensors` without the need for any additional programming. These measurements can be assessed by the size of the blue (ultrasound) and red (infrared) beacons, as well as the values displayed in the `leftSensor`, `frontSensor` and `rightSensor` boxes.

5   Conclusions and Future Extensions

This paper introduces an educational programming language architecture to teach students how to control and program simple behaviors with ROS-enabled robots. The architecture combines *ROS* APIs and libraries, such as `rosbridge` and `roslibjs`, with the *Snap!* visual programming language. The architecture targets students from primary education to high school, with the intent to increase both engagement and skills in STEM-related topics. The proposed approach introduces programming in the early stages of education, while still benefiting from the power of ROS, bypassing the understanding of complex language syntax and other programming paradigms.

The authors intend to further extend the reach of the programming architecture by integrating other robotic platforms. Although the architecture is currently prepared to publish and subscribe topics in any ROS-enabled robot, some features, such as the visual sprites and sensor readings, are still based on the *RobotCraft Bot v2.0* platform. As a future work, the authors foresee to implement a WebGL visualizer, moving from 2D animation to more realistic 3D graphics that, besides improving the visual attractiveness of the GUI, would broaden the range of applications and functionalities, namely moving from planar ground robots to aerial and aquatic solutions.

Acknowledgement. We sincerely thank the community for the contributions on the free and open-source frameworks adopted in this work, particularly: robotcraft.ingeniarius.pt, ros.org, robotwebtools.org and snap.berkeley.edu.

References

1. Benitti, F.B.V.: Exploring the educational potential of robotics in schools: a systematic review. Comput. Educ. **58**(3), 978–988 (2012)
2. Alimisis, D.: Educational robotics: open questions and new challenges. Themes Sci. Technol. Educ. **6**(1), 63–71 (2013)
3. Kim, S.H., Jeon, J.W.: Programming Lego mindstorms NXT with visual programming. In IEEE International Conference on Control, Automation and Systems (2007)
4. Michieletto, S., Ghidoni, S., Pagello, E.: Why teach robotics using ROS? J. Autom. Mobile Robot. Int. Syst. **8**(1), 60–68 (2014)
5. Quigley, M., Conley, K., Gerkey, B., Faust, J., et al.: ROS: an open-source robot operating system. In: ICRA Workshop on Open Source Software, vol. 3, no. 3.2, (2009)
6. Mavrotheris, E., Meletiou-Mavrotheris, M.: SMASH: an innovative training approach for parent education in mathematics and science. In: Proceedings of the 6th Hellenic Conference for Information and Communication Technologies in Education, pp. 349–356 (2008)
7. Joseph, L.: Mastering ROS for Robotics Programming. Packt Publishing, Birmingham (2015)
8. Noori, F.M., Portugal, D., Rocha, R.P., Couceiro, M.S.: On 3D simulators for multi-robot systems in ROS: MORSE or Gazebo?. In: Proceedings of the 15th IEEE International Symposium on Safety, Security, and Rescue Robotics (SSRR), Shanghai, China (2017)
9. Araújo, A., Portugal, D., Couceiro, M.S., Rocha, R.P.: Integrating arduino-based educational mobile robots in ROS. J. Int. Robot. Syst. **77**(2), 281–298 (2015)
10. Couceiro, M.S., Figueiredo, C.M., Luz, J.M.A., Ferreira, N.M., Rocha, R.P.: A low-cost educational platform for swarm robotics. Int. J. Robots, Educ. Art **2**(1), 1–15 (2012)
11. Kuipers, M.: Localization with the iRobot create. In: Proceedings of the 47th Annual Southeast Regional Conference, p. 33. ACM (2009)
12. Zaman, S., Slany, W., Steinbauer, G.: ROS-based mapping, localization and autonomous navigation using a pioneer 3-DX robot and their relevant issues. In: IEEE Saudi International Electronics, Communications and Photonics Conference, SIECPC 2011 (2011)
13. Crick, C., Jay, G., Osentoski, S., Pitzer, B., Jenkins, O.C.: Rosbridge: ROS for non-ROS users. In: Robotics Research, pp. 493–504. Springer, Heidelberg (2017)
14. Toris, R., Kammerl, J., Lu, D.V., et al.: Robot web tools: efficient messaging for cloud robotics. In: Proceedings of the 2015 IEEE/RSJ International Conference on Intelligent Robots and Systems (IROS), pp. 4530–4537. IEEE (2015)
15. Bellas, F., Naya, M., Varela, G., et al.: The Robobo project: bringing educational robotics closer to real-world applications. In: International Conference on Robotics and Education, RiE 2017, pp. 226–237. Springer (2017)

16. Zubrycki, I., Kolesiński, M., Granosik, G.: Graphical programming interface for enabling non-technical professionals to program robots and internet-of-things devices. In: International Work-Conference on Artificial Neural Networks, pp. 620–631. Springer (2017)
17. Yumi, N., Yuki, S., Tetsuya, O.: An effective visual programming tool for learning and using robotics middleware. Proceedings of the 2016 IEEE/SICE International Symposium on System Integration, Sapporo, Japan (2016)
18. Weintrop, D., Wilensky, U.: To block or not to block, that is the question: students' perceptions of blocks-based programming. Proceedings of the 14th International Conference on Interaction Design and Children, Boston, Massachusetts, pp. 199–208 (2015)

Real-Time Matlab-Simulink-Lego EV3 Framework for Teaching Robotics Subjects

Nicolás Montés[1]([✉]), Nuria Rosillo[1], Marta Covadonga Mora[2], and Lucia Hilario[1]

[1] Department of Mathematics, Physics and Technological Sciences, Universidad Cardenal Herrera-CEU, CEU Universities, C/San Bartolome 55, CP 46115 Alfara del Patriarca (Valencia), Spain
{nicolas.montes,nrosillo,luciah}@uchceu.es
[2] Department of Mechanical Engineering and Construction, Universitat Jaume I, Avda. de Vicent Sos Baynat s/n, Castellón de la Plana, Spain
mmora@uji.es

Abstract. This work develops a new educational platform based on the Matlab-Simulink package for the teaching of robotics using the Lego EV3 platform. The majority of Lego platforms used in the literature are NXT platforms, as the EV3 platform is relatively new (January 2013). Moreover, in contrast to previous Lego robot versions, this platform allows to develop a real-time framework to teach Robotics subjects. The framework is based in Matlab, the most widely used programming environment with LEGO Mindstorms, and employs the package provided by MathWorks. The proposed framework is tested here for a new motion planning algorithm, where the user can interact with the environment and the robot in real-time via a web camera. To the authors' knowledge, this is the first time a real-time application, capable to interact with the student, is developed to teach Robotics with EV3 platforms.

Keywords: Lego · Educational platform · Motion planning

1 Introduction

Over the last few years, in most universities around the world, specifically those offering Engineering degrees, low cost platforms have been used to improve the teaching experience in several subjects, in particular those related to Robotics, but also for developing other kinds of applications focused on data processing, advanced measuring and communication, amongst other possibilities. Furthermore, these low-cost platforms have such high capabilities that they can even be employed in some research developments. Most useful low-cost platform are [1]: Arduino, the Raspberry Pi, Kinect, etc. In the case of mobile robots, in the last years it has been possible to acquire low cost mobile robotic platforms as Adept [3], Moway [4], Epuck [2] and LEGO Mindstorms [5]. These low-cost

© Springer Nature Switzerland AG 2019
W. Lepuschitz et al. (Eds.): RiE 2018, AISC 829, pp. 230–240, 2019.
https://doi.org/10.1007/978-3-319-97085-1_23

platforms do not offer the same accuracy as industrial robots but they are valid for educational activities and even research and experimental tests.

Although all low cost platforms are useful, the role that LEGO Mindstorm has played in college engineering education over the last 15 years is striking. The main reason is that Lego Mindstorms series of kits contain software and hardware to create customizable, programmable robots. They include an intelligent brick with a computer that controls the system, a set of modular sensors and motors, and Lego parts from the Technic line in order to create the mechanical systems, which allows to compose many different robotic systems. Lego robotics is fundamentally a constructivist tool, with students leveraging their knowledge and experience to solve a real-world problem and to consistently question and challenge that knowledge as they develop their solution [6].

2   LEGO Mindstorms Platforms

Fig. 1. LEGO Mindstorms platforms. RCX (left), NXT (Middle) and EV3 (right).

1. **RCX: Robotic Command Explorer**. Inside of the RCX there is an 8-bit Hitachi H8/3292 microcomputer with 32 KB RAM and 16 KB flash memory. Instructions are executed with a speed of 16 MHz. Instead of USB, RCX uses infrared signals to communicate with a PC or another RCX.
2. **NXT: The Next Generation**. The first NXT version was launched in 2006 and the NXT 2.0 was launched in 2009. Inside of NXT there is a 32-bit AT91SAM7S256 microcontroller with 64 KB RAM and 256 KB flash memory. NXT uses a coprocessor, an Atmel 8-Bit ATmega48, to control the motors. The execution speed is 48 MHz. NXT uses Bluetooth communication.
3. **EV3: The Evolution (3rd Generation)**. Available since January 2013, EV3 has a 32-bit ARM9 processor. The size of the EV3 RAM has increased from 64 KB to 64 MB and the size of the flash memory has increased to 16 MB. The system clock speed has increased to 300 MHz, which is 6 times faster than that of NXT. A new feature of EV3 over NXT and RCX is the Wi-Fi connection to a network. It requires a wireless adapter, also called a dongle. EV3 sup-ports Linux operating system and could be programmed in C++, among other languages, changing the firmware (Fig. 1).

3   LEGO Mindstorms for Education

LEGO Mindstorms is used for teaching in many different subjects. The literature offers numerous research results about it. As the present paper is focused in coding and robotics subjects, detailed state-of-the-art in this specific areas is provided in the following subsections.

3.1   Teaching Code

Regarding code teaching, in [7] was investigated how engineering faculty student achievements were affected by robotics technology when learning computer programming algorithm logics by means of LEGO Mindstorms EV3. The statistical results showed in [7] demonstrated that the Lego robot usage raised student achievements in an introductory computer programming course. LEGO Mindstorms has been used to teach the most common programming languages such as C, Java, ADA, Phyton, Labview, and Matlab/Simulink. In some cases the goal is directly teaching code. In other cases, the language code is supposed to be known and used in class to reinforce the code knowledge.

MATLAB is the most widely used program with LEGO Mindstorms, see [8–17]. Although not free, MATLAB is extensively used in universities thanks to the student licenses. MathWorks has developed a free package for Matlab-Simulink called "Simulink Support Package for LEGO MINDSTORMS". It allows to program a LEGO robot by means of Simulink blocks with LEGO code.

3.2   Teaching Robotics

Regarding Robotics teaching, many experiences demonstrate the utility of LEGO platforms in different levels of education but mainly in higher education. For instance, in [18] a LEGO Mindstorms NXT was used to guide the students of the fifth year of an integrated Master's program in understanding the concept of localization in mobile robotics, using the Extended Kalman Filter (EKF). In [19] a set of projects are described that put into practice Artificial Intelligence techniques using LEGO Mind-storms NXT in an undergraduate Computer Science degree. They developed reactive and deliberative agents, rule-based systems, graph search algorithms, and planning. In [20] LEGO NXT Mindstorms was used with senior undergraduate engineering students, the 2nd Summer School on Mechatronics, to teach the basics of mobile robot control, teleoperation, line-tracking, fuzzy logic control techniques and reactive navigation. In [12] an educational hands-on project subsystems was proposed for learning guidance and control. A Lego Mindstorms NXT, a web camera, and tractable tools were used for searching and mapping an obstacle in an indoor environment. Lego robot visual tracking was implemented using color marker detection and an EKF. In [21], LEGO Mindstorms NXT robots were used in a course of the Foundation of Industrial Robotics for the Master Degree in Automation Engineering at the Engineering School of the University of Bologna. The goals were, on the one hand, to drive the students in the acquisition of practical knowledge regarding mobile robotics and, on the other hand, to teach them how to write efficient code for real time control of automatic machines and robots. The course used some usual tasks in the teaching of robotics like car parking, sumo robot, mobile gripper and pallet transportation.

3.3 Teaching Particular Robot Models

LEGO platforms have also been used for teaching robotic modeling. In [22] LEGO Mindstorms was used in laboratory practices with manipulators, where different manipulator models were constructed, such as an anthropomorphic robot, a Scara robot, and even a Cartesian robot used to write. This lab experiences were carried out in the first of the two years of the M.Sc. degree in Computer and Mechatronic Engineering.

LEGO RCX Mindstorms kits were used in [23] for teaching traction and ballasting of tractors in the Agricultural Equipment and Machinery course at the University of Missouri.

In [24] LEGO Mindstorms NXT was used to design a model of an armored car, named LEGOPardo, with the aim of presenting the main theoretical concepts of the course in such a way that cadets could establish a correspondence between the theory and its practical effects.

In [15] a Mechatronic Demonstrator was proposed that used the LEGO NXT Platform. This demonstrator, named humanoid robot, was suitable for studying the key elements of mechatronic systems: mechanical system dynamics, sensors, computer interfacing, and application development.

In [14] a LEGO Mindstorms NXT ballbot was employed to teach linear controllers with parameter variations, not only for undergraduate, but also for postgraduate students. The ballbot is a skinny, agile, and omnidirectional robot that balances on, and is propelled by, a single spherical wheel. The ballbot dynamics is based on the spherical inverted pendulum, and it is an interesting object of study for students, engineers, and researchers in automatic control. Inverted pendulum was also considered as at size scenario in [11] to outline both the theoretical and practical aspects of the Model Predictive Control theory.

In [25,26] LEGO Mindstorms is used to teach a self-stabilized bicycle, which is a complex model-based control design founded also in an inverted pendulum. In [27] a snail robot was constructed based on EV3 LEGO Mindstorms that is capable of climbing up a ladder.

3.4 Goal of the Paper

The goal of the paper is to present a new educational platform based on the Matlab-Simulink package to teach Robotics. Most of the Lego platforms used in the literature are NXT platforms. The main reason is that the EV3 platform is relatively new (January 2013). The advantage with respect to previous Lego robot versions is that this platform allows to develop a real-time framework for teaching robotics subjects. The framework is based in Matlab, the most widely used program with LEGO Mindstorms, using the package provided by MathWorks.

The paper is organized as follows. Section 4 explains the construction of a real-time Matlab-Simulink- EV3 framework. Section 5 tests this platform with a new motion planning algorithm where the user can interact with the environment in real-time. Conclusions are drawn in Sect. 6.

4   Matlab-Simulink-Lego EV3 for Teaching Robotics

In MATLAB® and Simulink® environments, free libraries can be downloaded for programming and communicating with Lego NXT and EV3 robots. Figure 2 shows the available blocks in the Simulink® library. The blocks in this tool are simpler than those available in the LEGO software.

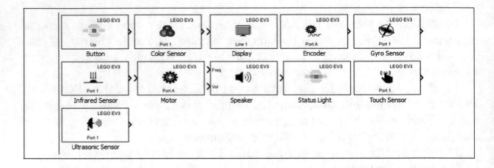

Fig. 2. Simulink available blocks for programming Lego EV3 robots

There are blocks for modelling colour, gyro, ultrasonic, infrared touch and ultrasonic sensors. For the control of the device, the motor and the encoder (rotation sensor) blocks have been used. A single block is available for motor modelling, without distinguishing between large or medium motor. The power value is an input to the block (with values between −100 and 100, as in the case of the LEGO software) and the connecting port must be determined in the block settings. The encoder allows reading the angle travelled in degrees.

This library is very easy to use, as it works as common Simulink blocks. When the code is finished, Simulink translated to C++ and transfers it to the Lego Mindstorms EV3 through Wi-fi connection. Other Simulink blocks can be also employed in the Lego robot, with a similar functioning.

Regarding the code execution in the LEGO EV3 robot, some options are available to the programmer. One of them, used in the present work, is the Simulink® External Mode. In real-time the External Mode, the Simulink® Coder™ dynamically links the generated algorithm code with the I/O driver code generated from the I/O blocks. The resulting executable file runs in the operating system kernel mode on the host computer and exchanges parameter data with Simulink® via a shared memory interface.

Parameters changed in the Simulink® block diagram are automatically updated in the real-time application thanks to the Simulink® external mode. The external mode executable is fully synchronized with the real-time clock.

4.1 Linking Matlab-Simulink-EV3

Simulink® Model

In the proposed Matlab-Simulink-EV3 framework, a Simulink template is created for modelling the system, as displayed in Fig. 3.

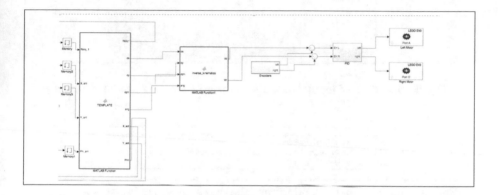

Fig. 3. Simulink template

The following blocks can be distinguished in the model: Template, PID Controller, Engines, Encoders, and Inverse Kinematics (IK). The robot inverse kinematics is coded in the last block in order to obtain the wheels velocities.

The inputs in the PID Controller block are the errors between the wheels velocities from the IK block and the velocities measured by the encoders, and the output is the power to supply to the engine.

The Template block is a Matlab function block, where motion planning algorithms can be coded to be tested in the LEGO Robot. This block has three I/O variables: Inputs, Outputs, and Parameters. These variables are defined in the "Port and Data Manager" window of the Matlab function, shown in Fig. 4. The first two variables are the Inputs and Outputs in the Simulink model but Parameters relate to the Matlab Workspace. Parameters can be selected as "tunable", meaning that can be modified in Real Time through the Matlab Workspace, or "non tunable" if they are constant and set when the template is compiled.

MATLAB® Environment

In the MATLAB environment, 'tunable' parameters must be defined and initialized first. For instance, for the 'tunable' parameter POS_A the following code is required to define it and initialize it:

Fig. 4. Definition of inputs, outputs and parameters in a Matlab function

```
POS_A=Simulink.Parameter;
POS_A.CoderInfo.StorageClass='ExportedGlobal';
POS_A.Value=[0 0];
```

Besides the parameter definition, the next sentence must be loaded in Simulink:

```
set_param('Plantilla_lego','SimulationCommand','Update');
```

Notice that Plantilla_lego is the name of the Simulink file. Parameters must be defined and initialized only once. However, each time a parameter value is changed in Matlab, the value must be reloaded in Simulink. A schema of the proposed parameter update system is depicted in Fig. 5.

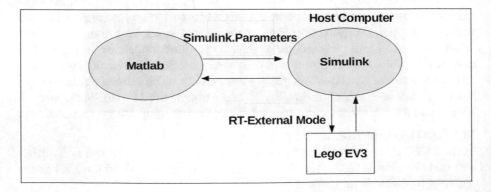

Fig. 5. Linking Matlab-Simulink-EV3

5   Testing the Platform with a New Motion Planning Algorithm

The framework proposed in the present paper has been tested with a new motion planning algorithm called PGD, see [12]. The experimental setup is depicted in Fig. 6. A $2 \times 2\,\mathrm{m}^2$ white square table is located on the floor and a web camera is located on the ceiling, as shown in Fig. 6 (left). The camera is connected via USB to the PC.

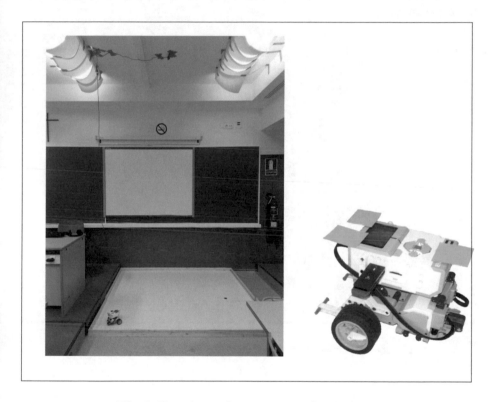

Fig. 6. Experimental setup to test the platform.

The Lego Robot is equipped with three blue points that locate its position and orientation via images captured by the camera, Fig. 6 (right). The goal point is a red disk. The new algorithm is computed in the template Simulink (MATLAB function block) and the robot position and orientation, as well as the goal, are defined as 'tunable parameters'. In the Matlab environment, two different files are developed. The first one defines and initializes the parameters and the camera. The second one takes snapshots in real-time with the camera, computes the current position and orientation of the robot and the goal and updates the parameters in Simulink. The execution sequence is:

1. Execute the file that defines and initializes the parameters and the camera in the MATLAB environment.
2. Execute the Simulink template in External Mode.
3. Execute the file that updates the current position and orientation of the robot and the goal.

Figure 7 shows four snapshots about the results. As we can see in this experiment, the user can interact with the framework in real-time, in this case changing the goal. There is another experiment changing the robot position. To watch all the experiments, visit [28].

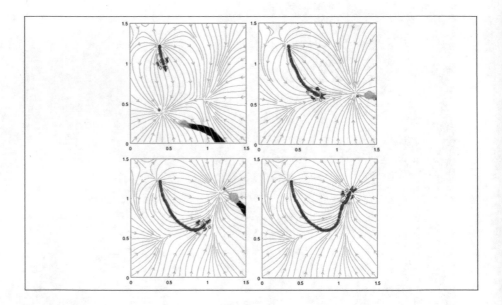

Fig. 7. Snapshots of an experiment for testing the framework.

6   Conclusions

The present paper develops a new educational platform based on the Matlab-Simulink package for robotics teaching using Lego EV3. For the first time, this platform is used in teaching applications, since it is relatively new. The main advantage offered in contrast to previous versions is its possible use in real-time. This property allows the implementation of real-time robotics applications in a simple and efficient way, which motivates students and lecturers in the teaching of robotics subjects.

The framework has been developed in Matlab-Simulink using the free package provided by MathWorks. To demonstrate the properties of the proposed framework, a new motion planning algorithm has been implemented in a real mobile Lego robot where the user can interact in real-time with the environment and the robot.

The results of the present paper are under patented process.

References

1. Irigoyen, E., Larzabal, E., Priego, R.: Low-cost platforms used in control education: an educational case Study. In: 10th IFAC Symposium Advances in Control Education, pp. 256–261 (2013)
2. Epuck, 2014. e-puck education robot. http://www.e-puck.org. Accessed 26 Oct 2017
3. ADEPT, 2014. Adept Mobile Robots. http://activrobots.com. Accessed 26 Oct 2017
4. Moway, 2017. Moway Robots. http://moway-robot.com. Accessed 26 Oct 2017
5. Mindstorm LEGO, 2017. LEGO Mindstorms. http://mindstorms.lego.com. Accessed 26 Oct 2017
6. Danahy, E., Wang, E., Brockman, J., Carberry, A., Shapira, B.: LEGO based robotics in higher education: 15 years of students creativity. Int. J. Adv. Robot. Syst. **11**(2), 157–172 (2014)
7. Ozurcun, N.Ç., Bicen, H.: Does the inclusion of robots affect engineering students achievement in computer programming courses. J. Math. Sci. Technol. Educ. **13**, 4779–4787 (2017)
8. Behrens, A., Atorf, L., Schwann, R., Neumann, B., Schnitzler, R., Balle, J., Herold, T., Telle, A., Noll, T.G., Ameyer, K., Aach, T.: Matlab meets LEGO mindstorms- a freshman introduction course into practical engineering. IEEE Trans. Educ. **53**(2), 306–317 (2010)
9. Kim, Y.: Control system lab using a LEGO mindstorms NXT motor system. IEEE Trans. Educ. **54**(3), 453–461 (2011)
10. Klassner, F., Peyton-Jones, J.C., Lehmer, K.: Genetic algorithms with Lego Mindstorms and Matlab. In: Proceedings of the Twenty-Fifth International Florida Artificial Intelligence Research Society Conference, pp. 312–317 (2012)
11. Cruz-Martín, A., Fernández-Madrigal, J.A., Galindo, C., Gonzalez-Jimenez, J., Stock-mans-Daou, C., Blanco-Claraco, J.L.: A LEGO Mindstorms NXT approach for teaching at data acquisition, control systems engineering and real-time systems undergraduate courses. Comput. Educ. **59**, 974–989 (2012)
12. Canale, M., Casale-Brunet, S.: A multidisciplinary approach for model predictive control education: a LEGO Mindstorms NXT-based framework. Int. J. Control, Autom. Syst. **12**(5), 1030–1039 (2014)
13. Kim, S., Oh, H., Choi, J., Tsourdos, A.: Using hands-on project with Lego Mindstorms in a graduate course. Int. J. Eng. Educ. **30**(2), 458–470 (2014)
14. Sood, V., Wadoo, S.: Establishing a cost effective embedded control and robotics program: delay based wireless feedback control using LEGOs. In: IEEE Integrated STEM Education Conference, pp. 210–215 (2015)
15. Garcia-Garcia, R.A., Arias-Montiel, M.: Linear controllers for the NXT ballbot with parameter variations using linear matrix inequalities. IEEE Control Syst. **30**(6), 121–136 (2016)
16. Savu, D., Sandru, L.A., Crainic, M.F., Moldovan, C., Dolga, V., Preitl, S.: Multiple methods of data acquisition for a LEGO NXT 2 mobile robot: the use of a second NXT 2 hardware platform. In: Proceedings of the 5th International Conference on Mechatronics and Control in Engineering, pp. 97–102 (2016)

17. Ding, J., Li, Z., Pan, T.: Control system teaching and experiment using LEGO Mindstorms NXT robot. Int. J. Inf. Educ. Technol. **7**(4), 309–317 (2017)
18. Serrano, V., Thompson, M., Tsakalis, K.: Learning multivariable controller design: a hands-on approach with lego robotic arm. In: Advances in Automation and Robotic in Latin America. Lecture Notes in Networks and Systems, pp. 271–278 (2017)
19. Pinto, M., Moreira, A.P., Matos, A.: Localization of mobile robots using an extended Kalman filter in a LEGO NXT. IEEE Trans. Educ. **55**(1), 135–145 (2012)
20. Cuellar, M.P., Pegalajar, M.C.: Design and implementation of intelligent systems with LEGO Mindstorm for undergraduate computer engineers. Comput. Appl. Eng. Educ. **22**(1), 53–166 (2011)
21. Gomez de Gabriel, J.M., Mandow, A., Fernandez-Lozano, J., Garca-Cerezo, A.J.: Using LEGO NXT mobile robots with LABVIEW for undergraduate Courses on mechatronics. IEEE Trans. Educ. **54**(1), 41–47 (2011)
22. Grandi, R., Falconi, R., Melchiori, C.: Robotic competitions: teaching robotics and real-time programming with LEGO mindstorms. In: Proceedings of the 19th World Congress The International Federation of Automatic Control, vol. 47, no. 3, pp. 10598–10603 (2014)
23. Indri, M., Lazzero, I., Bona, B.: Robotics education: proposals for laboratory practices about manipulators. In: IEEE International Conference on Emerging Technologies and Factory Automation (2013)
24. Bulent Koc, A., Liu, B.: Demonstrating tractor rollover stability using Lego Mindstorms and smartphones. J. Agric. Syst. Technol. Manag. **24**, 1–11 (2013)
25. Basso, M., Innocenti, G.: Lego-bike: a challenging robotic lab project to illustrate rapid prototyping in the mindstorms/simulink integrated platform. J. Comput. Appl. Eng. Educ. **23**(6), 947–958 (2015)
26. Budaciu, C., Apostol, L.D.: Dynamic analysis and control of Lego Mindstorms NXT bicycle. In: IEEE International Conference on System Theory, Control and Computing, pp. 145–149 (2016)
27. Vokorokos, L., Mihalov, J., Chovancova, E.: Potential of LEGO EV3 mobile robots. Acta Electrotechnica et Informtica. **15**(2), 31–34 (2015)
28. https://www.youtube.com/watch?v=LC_kFZPmOH0

IDEE: A Visual Programming Environment to Teach Physics Through Robotics in Secondary Schools

Samantha Orlando, Elena Gaudioso[✉], and Félix de la Paz

Department of Artificial Intelligence, Universidad Nacional de Educación
a Distancia (UNED), 28040 Madrid, Spain
sorlando1@alumno.uned.es, {elena,delapaz}@dia.uned.es

Abstract. One of the main difficulties that students face in physic courses in secondary school is the lack of interest about theoretical problems. Educational robotics have the potential to be used as powerful educational tools to motivate students. Nevertheless, learning activities with robotics kits are usually isolated from official curriculum. In fact, there is usually no evaluation about student learning outcomes in these activities. In this paper, we present IDEE, a learning environment that, through a visual programming environment and a robotic kit, proposes to the students a set of learning activities following a scientific procedure approach. All the interaction data about the performance of students is stored in a database. This information makes it possible to adapt the statements of the problems to the profiles of the students, while allowing support to teachers in the assessment of the learning process.

Keywords: Educational robotics · Physics · Secondary school
STEM · Constructivism

1 Introduction

Following the constructionism paradigm [8], students should construct their knowledge under the guidance of the teachers. To encourage students' active learning, besides theoretical content, it is required to provide practical exercises and up-to-date tools in laboratories. This is especially true in scientific subjects, such as physics.

Nowdays, educational robotics area has provided new environments and learning tools to promote constructivism education [2]. Due to the interest that students show in the interaction with a robot [6], educational robotics favor cognitive processes and generate active and constructive learning [11]. In addition, robotics are considered a valuable educational resource to promote the natural development of knowledge in STEM (Science, Technology, Engineering and Mathematics) subjects in school [2].

Nevertheless, educational robotics is currently used predominantly in extracurricular classes and isolated from the main curriculum [7]. Thus, it is

© Springer Nature Switzerland AG 2019
W. Lepuschitz et al. (Eds.): RiE 2018, AISC 829, pp. 241–246, 2019.
https://doi.org/10.1007/978-3-319-97085-1_24

difficult for the students to establish the cognitive relationships between the theoretical concepts and the concepts involved in the activities done with robotics.

In addition, educational robotics are usually used in classroom making the students to interact with robots directly from the available artifacts (for example, using Scratch directly to control a Lego robot Kit). These approaches do not integrate the activities in a pedagogical way. To overcome this limitation, new learning environments are being proposed in order to integrate learning activities and interaction with robotics in an unique environment. For example, in [5], it is presented a robot-visual programming environment interface to teach C Programming. [1] presents a visual programming environment that includes instructions that enable students to build and play musical sequences.

However, these environments do not track the students' interactions in order to adapt the sequence of activities or to provide support to teachers in the assessment of the learning processes. In this paper, we present IDEE (Integrated Educational Didactic Environment), a learning environment that uses robotics to provide learning activities to 15-years old students in a physics course in secondary school. IDEE is a visual programming tool that, together with a robotic kit, provide learning activities following a scientific experimental method while tracking students' interactions within the learning environment.

The rest of the paper is organized as follows: Sect. 2 presents a general overview of IDEE. Next, Sect. 3, describes the design of learning activities in IDEE. Finally, Sect. 4 concludes the paper and describe some lines of future work.

2   General Overview of IDEE

IDEE is a learning environment that provides a framework to solve learning activities in physics through the use of robotics. Students interact with IDEE through a web-based interface. This interface provides the problem statement and a visual programming environment.

Students in IDEE solve each learning activity using the visual programming editor that is based on drag-and-drop blocks. The definition of each learning activity in IDEE is based on the scientific experimental method. Learning activities in IDEE are composed of the following elements: statement of the problem, blocks that will be necessary for its resolution, solution and students' profiles to which it is recommended. A more complete description of a learning activity in IDEE is described in Sect. 3.

The necessary blocks in IDEE have been created within the learning environment. In order to guide the students, the set of blocks proposed depends on each problem, presenting those that would be used in the solution. Once the students have finished the problem, they submit the solution and IDEE evaluate it. In addition, IDEE can show the correct answer if the student experiences difficulties.

Moreover, students are asked to use robotics as laboratory artifacts in order to test their hypothesis about the solution of an activity. Through the use of the robot, students can visualize their proposed solutions as often as they wish while observing the behavior of the robot. Figure 1 depicts a screen-shot of IDEE learning environment.

The architecture of IDEE is based on Django[1], a framework that allows the creation and management of Python applications. The interface of IDEE is implemented with Google's Blockly Visual Programming Environment [10], which is based on Scratch [9]. IDEE is connected with a LEGO MINDSTORMS EV3 through a Scratch's plug-in that provides a bluetooth connection to the robot. The general architecture of IDEE is shown in Fig. 2.

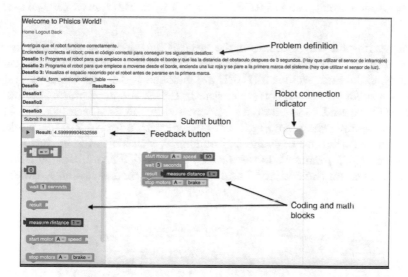

Fig. 1. General interface of IDEE. Notice that robot connection indicator changes from red to green to acknowledge the student when the robot is connected with IDEE.

In IDEE, students' evaluation is a fundamental tool to help teachers in supporting students in their learning process. This support should not be only based on final marks but in the whole learning process [3]. Depending on the performance of the students, IDEE maintains different types of profiles, namely, novice, intermediate and advanced. According to these profiles, IDEE presents different versions of the problem statement for each student. This adaptation is done by means of several predefined rules defined by a physics teacher.

To create and maintain the profiles, each learning activity in IDEE is related to the skills that each student must achieve (such as, knowledge about sensors in robotics or knowledge about magnitudes in physics). To determine the level of knowledge of each skill, IDEE presents the students a test for each learning activity to evaluate the background knowledge in the concepts involved in the activity. In addition, the answers that students submit in IDEE are evaluated

---

[1] https://www.djangoproject.com/.

244   S. Orlando et al.

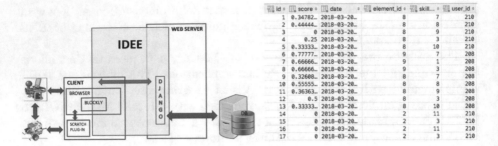

id	score	date	element_id	skill...	user_id
1	0.34782...	2018-03-20...	8	7	210
2	0.44444...	2018-03-20...	8	8	210
3	0	2018-03-20...	8	9	210
4	0.25	2018-03-20...	8	3	210
5	0.33333...	2018-03-20...	8	10	210
6	0.77777...	2018-03-20...	9	7	208
7	0.66666...	2018-03-20...	9	1	208
8	0.66666...	2018-03-20...	9	3	208
9	0.32608...	2018-03-20...	8	7	208
10	0.55555...	2018-03-20...	8	8	208
11	0.36363...	2018-03-20...	8	9	208
12	0.5	2018-03-20...	8	3	208
13	0.33333...	2018-03-20...	8	10	208
14	0	2018-03-20...	2	11	210
15	0	2018-03-20...	2	3	210
16	0	2018-03-20...	2	11	210
17	0	2018-03-20...	2	3	210

Fig. 2. General architecture of IDEE

Fig. 3. Table `ephysic-profile`.

and depending on this evaluation the level of knowledge of the skills involved in the activity is increased or decreased. The relationships between learning activities, skills and profiles in IDEE are encoded in the database (see Fig. 4). When a student submits a correct answer in IDEE, the percentage of knowledge on the skills related to the activity is increased in the database (see Fig. 3).

In the example shown in Fig. 3, the student with `user_id` 210 has worked on the activities (attribute `element_id`) number 8, y 2. This student has a 35% of knowledge in skill number 7 in the activity number 8. The student with `user_id` 208 has worked on the activity 8 and 9 and has reached a 33% of knowledge on skill 10.

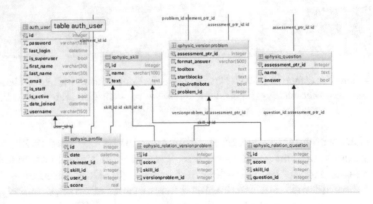

Fig. 4. Tables involved in the management of the relationship between learning activities, skills and profiles in IDEE.

3   Learning Activities in IDEE

The study of physical phenomenons in IDEE are based on the scientific experimental method: observation of the phenomenon, formulation of hypotheses, experimental verification, analysis of the data and conclusions. Each physical phenomenon in IDEE is composed by different experiments covering each of these steps. The objective is that the students feel that they are young researchers capable of responding to natural phenomena using mathematical language.

The study of every physical phenomenon in IDEE is composed of the following steps:

1. *Observation*: Initially the students observe the phenomenon to be studied thanks to a reproduction made by the robot. At this stage, the student do not interact neither with the robot nor with the system, but only observe the phenomenon.
2. *Assembly of the robot*: Students, with the aid of the teacher, have to assemble and test the robot to be used in the experiments.
3. *Experimental testing of hypotheses.* Following, IDEE presents several exercises that are related with the theoretical hypothesis about the phenomenon that it is being analyzed. Students solve the exercises through the visual programming environment and interact with the robot.
4. *Understanding of the physical magnitudes*: Once the phenomenon is observed, IDEE guides the student to the understanding of the physical magnitudes of each problem and the mathematical law that characterizes it.
5. *Conclusions.* Finally, students must draw conclusions on the phenomenon and understand the mathematical law that regulates it. In this phase, IDEE offers the student a problem that cannot be solved with the robot. Those exercises are presented considering the student's level of knowledge. The student would be able to answer the problem through algebraic tools. The goal is that the student can formalize the mathematical law that regulates the physical phenomenon studied in the activity.

4   Conclusions and Future Work

The main idea behind the creation of IDEE is to support students and teachers in the learning of physics concepts. This learning process is favored following the paradigm of an experimental scientific method.

Evaluation of the learning process is a useful instrument to help students to understand their mistakes and to strengthen the skills they should be lacking. IDEE stores all the data that results of the interaction of students with the system. In IDEE, the interaction and performance data of students while interacting are stored allowing the creation of different students profiles and the presentation of reports containing an analysis of the performance of students to the teachers.

One main line of future work is the definition and implementation of a dashboard for teachers to help them in the visualization of student's skills evolution by means of learning curves [4]. In addition, another line of future work is the implementation of a graphical interface for the teachers to define more learning activities in IDEE.

Additionally, to support students in the activities, the whole solution of each problem is currently shown. Nevertheless, this support should by more dynamic trying to help the student to find the solution by himself by only showing tips and not the complete solution.

Acknowledgment. This paper has been supported by the Italian Ministry "Ministero dell'Istruzione, dell'Universitá e della Ricerca", MIUR under law 488/2001.

References

1. Baratè, A., Formica, A., Ludovico, L., Malchiodi, D.: Fostering computational thinking in secondary school through music - an educational experience based on google blockly. In: Proceedings of the Ninth International Conference on Computer Supported Education (CSEDU 2017), pp. 117–124. SCITEPRESS (2017)
2. Barreto, F., Benitti, V.: Exploring the educational potential of robotics in schools: a systematic review. Comput. Educ. **58**(3), 978–988 (2012)
3. Gaudioso, E., Montero, M., Hernandez-del Olmo, F.: Supporting teachers in adaptive educational systems through predictive models: a proof of concept. Expert Syst. Appl. **39**(1), 621–625 (2012)
4. Koedinger, K.R., Stamper, J.C., McLaughlin, E.A., Nixon, T.: Using data-driven discovery of better student models to improve student learning. In: Lane, H.C., Yacef, K., Mostow, J., Pavlik, P. (eds.) Artificial Intelligence in Education: 16th International Conference, AIED 2013, Memphis, TN, USA, 9–13 July 2013. Proceedings, pp. 421–430. Springer, Heidelberg (2013)
5. Krishnamoorthy, S.P., Kapila, V.: Using a visual programming environment and custom robots to learn c programming and k-12 stem concepts. In: Proceedings of the 6th Annual Conference on Creativity and Fabrication in Education, FabLearn 2016, pp. 41–48. ACM, New York (2016)
6. Menegatti, E., Moro, M.: Educational robotics from high-school to master of science. In: Proceedings of the Conference on Simulation, Modeling and Programming for Autonomous Robots, pp. 639–648 (2010)
7. Ospennikova, E., Ershov, M., Iljin, I.: Educational robotics as an inovative educational technology. Procedia - Soc. Behav. Sci. **214**, 18–26 (2015)
8. Papert, S.: Mindstorms. Children, Computers and Powerful Ideas. Basic Books, New York (1980)
9. Resnick, M., Maloney, J., Monroy-Hernández, A., Rusk, N., Eastmond, E., Brennan, K., Millner, A., Rosenbaum, E., Silver, J., Silverman, B., Kafai, Y.: Scratch: programming for all. Commun. ACM **52**(11), 60–67 (2009)
10. Trower, J., Gray, J.: Blockly language creation and applications: visual programming for media computation and bluetooth robotics control. In: Proceedings of the 46th ACM Technical Symposium on Computer Science Education, SIGCSE 2015, p. 5. ACM (2015)
11. Usselman, M., Ryan, M., Rosen, J.H., Koval, J., Grossman, S., Newsome, N.A., Moreno, M.N.: Robotics in the core science classroom: benefits and challenges for curriculum development and implementation. In: Proceedings of the 2015 ASEE Annual Conference & Exposition, Seattle, Washington, pp. 26.1349.1–26.1349.16 (2015)

Cross Topics in Educational Robotics

Robotic Trains as an Educational and Therapeutic Tool for Autism Spectrum Disorder Intervention

Ahmad Yaser Alhaddad[1,2], Hifza Javed[3], Olcay Connor[5], Bilikis Banire[4], Dena Al Thani[4], and John-John Cabibihan[1(✉)]

[1] Department of Mechanical and Industrial Engineering, Qatar University, Doha 2713, Qatar
john.cabibihan@qu.edu.qa

[2] Department of Electronics, Information and Bioengineering, Politecnico di Milano, Piazza Leonardo da Vinci 32, 20133 Milan, Italy

[3] Biomedical Engineering Department, George Washington University, Washington, D.C. 20052, USA

[4] College of Science and Engineering, Hamad Bin Khalifa University, Doha 34110, Qatar

[5] Step by Step Center for Special Needs, Doha 47613, Qatar

Abstract. Social robots are emerging to become useful assistive tools for use in the therapy and education of children with Autism Spectrum Disorder (ASD). The nature of ASD causes its symptoms and manifestations to vary widely, resulting in a variety of robotic form factors that have been developed for this application. These robots vary in structure, shape, size, color, and function. In this paper, we propose a train-based model to be used as an educational and rehabilitation tool. We explore the responses from children with ASD in a special needs centre toward a variety of toys (e.g. ball, cymbals, truck) and robots (i.e. humanoid and a robotic seal), including trains. Experiments were conducted to verify whether or not trains have any advantage over other form factors and to extract any features of interest to the children. Results show that trains represent an appealing platform to a wider range of children with ASD. Additionally, results showed that simple features that can be easily incorporated into trains play a significant role in the interactions and could serve as reward mechanism.

Keywords: Autism · Social robots · Trains

1 Introduction

Characterized by lifelong difficulties in communication, social interaction, and behavior [1], Autism Spectrum Disorder (ASD) is a condition that is diagnosed during early childhood and affects neurodevelopment. The rate of autism among children has caused a lot of concern among parents, governments, and the medical and scientific communities. For example, the prevalence rate of autism has

© Springer Nature Switzerland AG 2019
W. Lepuschitz et al. (Eds.): RiE 2018, AISC 829, pp. 249–262, 2019.
https://doi.org/10.1007/978-3-319-97085-1_25

been reported to be affecting 1 out of 45 children in the United States [39]. Several technologies have been explored in supporting therapeutic and educational initiatives for children with ASD [15,29]. Many previous studies have shown that children on the spectrum have great in interest in technology, such as computer applications [18], virtual environments [26,36], and robots [8,19,20].

Robotics, among other type of technologies, seem to be more effective in teaching social and communication skills to children on the spectrum [34]. Emotion recognition is a crucial skill at which using social robots was proven to be helpful in supporting children with ASD recognizing emotions [28]. Some studies have shown that social robots were effective in training children with autism relying on imitation approaches [40]. Another study demonstrated that using music with a social robot served as a reinforcement in learning imitation skill [27].

When it comes to the elicitation of behaviors, robots have been successfully used to evoke many behaviors, such as gestures [10–12], joint attention, imitation, eye-contact [16,38], and interpersonal distance [17]. Because robots are less complex as compared to humans, children were found to be more responsive and less intimidated by robots [8]. There was a significant variation in the types of form factors considered that were either small or large in size while taking the appearance of either humans, animals, toys or others [20]. Due to the heterogeneity of ASD, the reactions of individuals with ASD toward the existing robotic designs have varied considerably, and so are their preferences.

There is a potential for new emerging robotic designs to be effective enough for therapy and to be used as learning tools. There have been reported evidences of the fascination between children with ASD and Thomas, the Tank Engine (Mattel, Inc, AR, USA) [21]. Furthermore, the train has been considered to teach children on the spectrum emotions in an animated series (i.e. The Transporters). In this study, we propose a model of a train to be considered as a robotic form and to be used as a learning and intervention tool. We also explored the feasibility of trains as a form factor among children with ASD based on their interactions with a variety of toys, including trains and social robots, as a part of pilot experiments for the proposed model.

This paper is organized as follows. Sect. 2 describes the proposed model. Sect. 3 describes the participants, stimuli, experimental setup, and analysis. Sect. 4 provides the results and discussions. Finally, Sect. 5 concludes the paper.

2   Smart Robotic Train: A New Intervention and Learning Tool

2.1   Motivation

Over the years, many social robot designs have been developed and tested for intervention sessions [8]. Those include humanoids, human-like robots, robotic balls, mobile robots, and animaloids. Among these designs, mobile robots seem to gain special interest among children with ASD. The fascination about mobile robots can be attributed to their ability to be autonomous, generate interesting and interactive interplays, and sustain attention through various ways [24]. Train toys have been reported to be popular among children with ASD [21]. The potential reported reasons behind this popularity are summarized below.

Trains generally have a strong appeal to children on the spectrum [2]. Individuals with ASD are in need for predictable patterns to make sense of the world around them. This need seems to be satisfied through the predictability provided by trains and their schedules.

Filled with many details, such as size, motion, sound, shape, and color, trains help children with ASD to fulfill the tendency to memorize and recite such details. This also helps in the cognitive development [3], as it satisfies their need to identify and organize things into groups. Children on the spectrum have been seen with their lined-up trains arranged according to colors or sizes. The distinguishable bold colors seem to help them identify different colors and characters, and help maintain longer engagements [25].

The mobility of trains has provided an added advantage to encourage children to move around while interacting with their toy train. This feature can help children develop their motor skills [31]. Children with autism have been reported to show interest in moving or spinning objects. The train wheels satisfy this interest and can serve as another feature to engage with the children.

It has been observed that children with autism often enjoy reenacting scenes, such as falling, crashing or smashing, using their toy trains. These scenes are usually displayed in Thomas and Friends videos, and they are very easy to reenact, and hence, improving their imitation skills. Thomas toy trains often generate reactions depending on the actions of the children [14], and that helps in perception development of the environment around them and how they should interact with it.

Trains, with their varying tracks, can be placed in various landscapes, such as cities, villages, caves, farms, and others. This diversity of potential sceneries helps the children learn more about the world around them and perform a lot of activities setting up these tracks with others [13].

2.2   The Proposed Model

The proposed design will account for the aggressive or self-injurious behaviors that are exhibited by children on the spectrum to ensure their safety [23,37]. This can be achieved by ensuring that the overall design does not include sharp edges and be made with soft materials (e.g. [4,5,9]) to mitigate any potential harm. For example, different biopolymer-based sensors could potentially be embedded within the soft material for monitoring and data collecting purposes [22,32].

The proposed design is modular and includes multiple and detachable carriages. Each carriage is meant to perform different activities and the overall pieces of the train can be appended to widen the scope of applications. The main piece of the proposed design will contain a face on its front to interact with the children. The face could be electronically or mechanically based and

Fig. 1. Proposed model for a smart robotic train [6].

should be able to exhibit the basic emotions and facial expressions. Further-more, this piece could be fitted with a bubbles-generating chimney mimicking the smoke produced by a real steam engine. The goal of this feature is to be used as a reward mechanism to evoke interest in the tool. For example, the train generates bubbles when the child is performing well in some of the tasks given during the therapy sessions.

The train could also include a container that holds cookies or candies that can also be used as rewarding mechanism. To satisfy their interest in technology and in robots, another carriage could contain a robotic arm that is capable of performing various activities. To further evoke their interest in the design, the train parts would be made colorful and its tracks could either be fixed or LED projected. The software aspect of the proposed design consists of various and easily downloadable activities. Each activity is meant to target a specific goal or elicit a specific wanted behavior. Figure 1 depicts some of the proposed features.

The objective of the experiments in this study is to investigate whether the train will be attractive to the children on the spectrum.

3   Methods

3.1   Participants

Ten English-speaking children aged 7 to 10 years (all were males) participated in this study. They have been diagnosed with mild to moderate autism and are attending the Step By Step Center for Special Needs in Doha, Qatar. The consent from the parents were secured by the center. The children were accompanied by either a teacher or a caregiver. The procedures for this work did not include invasive or potentially hazardous methods and were in accordance with the Code of Ethics of the World Medical Association (Declaration of Helsinki).

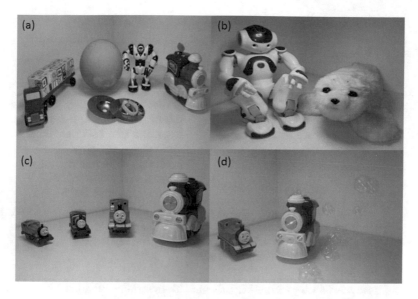

Fig. 2. The group of stimuli used during the experiments. (a) five different non-moving toys. (b) The social robots, Nao and Paro. (c) Thomas and Friends trains (left) and larger train (right). (d) The participant's favorite train from the previous stimuli group and bubbles train.

3.2   Stimuli

There were a total of 4 experiments, where different stimuli were used for each experiment. The details about the individual stimuli used are summarized below.

1. Five different toys were used. These were a rubber ball, two metal cymbals, a colourful plastic train, a small humanoid robot, and a wooden truck with wooden blocks pegged into its carrier that have alphabets and objects drawn on them (Fig. 2a).
2. Two interactive social robots were used (Fig. 2b). One was a Nao humanoid (SoftBank Robotics, Paris, France), and the other was a seal robot (PARO Robots U.S., Inc., IL, USA). The movements of Paro were autonomous and were limited to the built-in functions. The movements of Nao were initiated by the experimenter and were limited to basic activities (e.g. sit, stand up, dance).
3. Three Tank Engine toy trains were used. These included two blue trains of different sizes (Thomas train character) and one red train (James train character). These were used against a larger, multi-colored toy train (Fig. 2c).
4. The child's preferred train in Experiment 3 was used against a train identical to the multicolored train from Experiment 1 and Experiment 3, but emits bubbles (Fig. 2d).

Table 1. Experimental protocol and objective of each experiment

Expt #	Protocol	Objective
1	Present 5 different non-moving toys	To determine which toy is the most preferred
2	Remove all toys from Experiment 1 and replace with interactive social robots, such as Nao and Paro	To determine whether interactive robots, behaving autonomously, appeal to the child more than the non-moving toys from the preceding experiment, and to observe the nature of his/her interactions with them
3	Remove all toys from Experiment 2 and present the train from Experiment 1, and add different mechanical Thomas and Friends trains	To determine whether the interest in trains is limited to Thomas toys or extends to all trains
4	Remove all toys from Experiment 3 except the participant's favorite toy, and replace them with a train that automatically generates bubbles from its chimney once switched on	To determine whether bubbles can add to the appeal of a toy to verify their use as a reward mechanism

3.3   Experimental Setup

Procedure. There were 4 different experiments aiming at different goals (Table 1). Each experiment was around 6 min long. Experiment 1 is an unstructured play scenario. After obtaining the results of Experiment 1, the succeeding experiments explored animating the toys and the goal was to see the effects of these in the subjects. No instructions were given to the children, except encouragement to initiate interaction with different toys.

Monitoring Equipment. The children's interactions were monitored with four video cameras placed at the corners of the room. Four cameras (MyDlink DCS-931L, D-Link, Taipei, Taiwan) were mounted on four tripods. Care was taken in the setup of the equipment to ensure that it remained unobtrusive throughout the length of the experiment. The cameras were positioned to ensure that the children's activities were captured from different angles.

3.4   Analysis

An open-source video event-logging software (BORIS, version 3.12, Torino, Italy) was used for annotation. The user environment of the software was prepared with all the behaviors of interest. The analysis of videos was conducted by three observers after getting well-acquainted with the software.

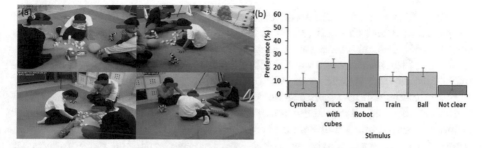

Fig. 3. The preferences for the toys in Experiment 1. (a) Child playing with the cubes of the truck. (b) Average rated preferences for each toy with the small robot scoring the highest.

The measured variables were divided to either state or point. A state variable was used to calculate the duration of a specific event while a point variable is used to calculate the frequency of occurrence. The measured variables are listed as follows:

- *Experiment duration*: state variable to declare the duration of an experiment.
- *Interaction duration*: state variable to declare the durations of interaction during an experiment.
- *Preference*: experiment-dependent point variable to indicate the preference of the child based on the given stimuli for each experiment. For example in Experiment 1, Q is pressed when the preference is the small robot. The deduction of preferences were based on either direct verbal communication, longest interaction duration or most preferred.
- *Unclear*: point variable is pressed once in case the preference implied by the child is not clear. Unclear selections occur when the child either prefers, selects more than one or neither of the stimuli.

4   Results and Discussions

Most of the children showed continuous interaction and engaging behaviors with the experimenter. The observed reactions varied differently across all the sessions, and the participants exhibited different responsiveness to a given stimulus. Some interactions only occurred after a prompt from the experimenter or caregiver while others were instantaneous and spontaneous. The preferences and the main features of interest have been observed and recorded based on post-analysis of the videos. The individual results and discussions are summarized below.

4.1   Experiment 1

In Experiment 1, five different toys were presented. Majority of the children preferred the small robot, followed by the truck with cubes and then the ball

(Fig. 3b). The small robot scored the highest (30%) followed by the truck with cubes (23.3%) and the ball (16.7%). The train scored 13.3% while the cymbals scored 10%. Unclear preference was around 6.7% of the participants.

Most of the interactions were limited to playing with the toys without standing up or moving around, and without showing high level of excitement. The majority of the children appeared interested to play with cubes that comes with the truck (Fig. 3a). They spent some time in picking the cubes while trying to name the shapes and numbers on them with the experimenter. When prompted, some of the children played with the ball together with the experimenter or caregiver. The sound and reflection of the cymbals piqued the interest of two children while three liked the features on the train, such as the colors and wheels. Three participants enjoyed interacting with the small humanoid robot the most while one avoided approaching it and kept his distance. This reaction could be attributed to the human-like appearance of the small robot. One participant did not show any interest in most of the toys in this experiment and ended up ignoring the toys.

The interest in technology, especially to robots, is evident in Experiment 1. This cannot be generalized across all the participants and all individual with ASD due to varying degree in their reactions to the same stimulus (e.g. small humanoid robot) and due to the small sample involved in our experiments. Some of the features and characteristics of existing toys (e.g. cubes on the truck) seem to still get the interest of the children on the spectrum and should be considered in any new robotic designs.

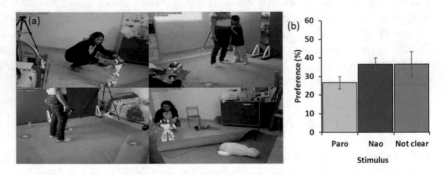

Fig. 4. The preferences for the social robots in Experiment 2. (a) Child crying and refusing to approach Nao. (b) Average rated preferences for each social robot. Less than half of the participants liked Nao more, while a quarter preferred Paro more. The remaining children did not like either of these robots.

4.2   Experiment 2

In Experiment 2, two interactive social robots were presented (Fig. 4a). Nao scored higher than Paro, 36.7% and 26.6%, respectively (Fig. 4b). 36.7% of the participants showed either refusal to interact or no definitive selection on the preference.

The observations can be divided to three groups. The first group of children interacted right away and showed positive reactions to the robots. They imitated or gave instructions to Nao and they played with Paro. The second group hesitated to interact quickly. They started by observing the movements of the robots, and they then began to approach the robots slowly to initiate the interactions. The last group refused to interact with either of the robots as they seemed afraid of interacting or even approaching the robots. Two children reacted with immediate anxiety upon the introduction of the robots and demanded the robots to be removed from the room (Fig. 4a).

The increased interactions and higher levels of excitement were clear in Experiment 2 among some of the participants. Part of that could be attributed to the novelty effect of the presented social robots [33]. Interestingly, the size of Nao being larger as compared to the smaller robot in Experiment 1 played an important role in altering some of the reactions negatively. This could imply that some children on the spectrum could feel more comfortable dealing with robots relatively smaller than themselves. Perhaps the smaller size gives them more sense of control over the presented stimuli. These negative reactions could also be attributed to lifelikeness of the presented social robots (i.e. human-like or animal-like).

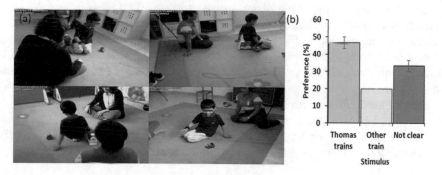

Fig. 5. The preferences for the trains in Experiment 3. (a) A child showing interest in the moving train. (b) Average rated preferences for the trains. The majority of children have selected Thomas and Friends trains. Around one-third of the children were not clear on their preference as they have interacted almost equally with all the trains without implying their preferences.

4.3   Experiment 3

In Experiment 3, three different Thomas and Friends trains and one bigger train were presented. Around 46.7% of the participants showed interest in Thomas trains and 20% showed interest in the other train. The rest were not clear on their preferences (Fig. 5b).

The children were more excited in this experiment as compared to Experiment 1 and Experiment 2. Some children showed more excitement and more movements when some of the trains were powered on (Fig. 5a). These reactions support the idea that the implementation of simple features could serve as reward mechanisms. Some recognized Thomas Trains and started re-enacting crashing scenes while mimicking the sound of a train. One child did not seem to show the same level of interest and excitement as compared to others.

Familiarity with a presented stimuli (e.g. Thomas Trains) seems to play a role in making interactions more fluid and spontaneous. Researchers in social and educational robots could exploit this aspect in promoting their designs to achieve higher effectiveness for the intended goals and purposes. This could be achieved by accompanying their developed robots with interactive videos and books exhibiting their robots in action. The exposure to such media prior to the presentation of the actual stimuli could help achieve better outcomes and mitigate any potential negative reactions.

4.4   Experiment 4

In this experiment, the bubble-generating train and the child's favorite train in Experiment 3 were presented (Fig. 6b). Almost all the participants showed high interest towards the train with bubbles (93.3%).

Most participants were immediately attracted to the bubbles, clapping or jumping in excitement, attempting to catch them. The interaction durations increased dramatically during this experiment. Children even showed more interactions with higher level of excitement depicted by the increased physical interactions, laughters, and movements (Fig. 6a). These observations support the idea of using bubbles as a reward mechanism in our proposed model. Some showed curiosity about the train and its features that they ended up carrying it while walking around. Few participants seemed to be wondering why the other big train was not generating any bubbles.

Similar outcomes for Experiment 4 could have occurred if some of the other toys presented in Experiment 1 were equipped with similar feature (i.e. bubbles

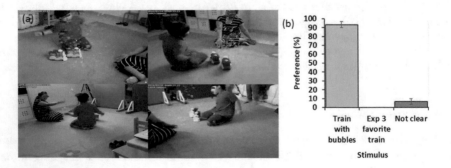

Fig. 6. The preferences for the trains in Experiment 4. (a) Child interacting with the bubbles generating train. (b) Average rated preferences for the trains. Almost all the children preferred the train with bubbles.

generating mechanism). However, the train with its carriages makes it easier to attach or de-attach new features (cf. Fig. 1). This provides more flexibility in changing the reward mechanism type and the activities by simply changing the accompanying carriages. Trains as a robotic platform provide more flexibility and adaptability to different scenarios and to various targeted end-users.

5 Conclusions

The results suggest diverse preferences among the participants. The most positive reactions were observed during the sessions with the train with bubbles. On the other hand, the instances where interactions were more difficult occurred during the sessions with the social robots, especially with Nao. While the humanoid robots have been reported to be a preferred candidate for imitation and eye-contact [35], the life-likeness of their appearance, relatively large sizes or sudden motions might have been a contributing negative triggers for this difficulty. There have been some instances of aggressive behaviors towards the social robots, such as pushing Nao and jumping on Paro. This could suggest that social robots that resembles the appearances of human or animal, to some degree, might not be positively perceived equally by children on the spectrum. However, repeated exposure and multiple sessions might alter these reactions overtime [30].

The train form factor has a strong potential to be considered in ASD therapy. Trains represent an appealing and neutral platform. With the addition of minor features, such as targeted activities, mobility and bubbles generation, trains could be used as educational, interactive and entertaining tool. These findings are important and will be used as inputs for the development of more interactive robotic trains, especially when taken in the context of product development (cf. [6]). Future work involves detecting the children's physiological responses when various stimuli are presented to them using available sensing technologies [7].

Acknowledgments. The work is supported by a research grant from Qatar University under the grant QUST-1-CENG-2018-7. The statements made herein are solely the responsibility of the authors.

References

1. Baio, J.: Prevalence of autism spectrum disorders: autism and developmental disabilities monitoring network, 14 sites, United States, 2008. Morb. Mortal. Wkly. Rep. Surveill. Summ. **61**(3), 1–19 (2012). Centers for Disease Control and Prevention

2. Baron-Cohen, S.: The extreme male brain theory of autism. Trends Cognit. Sci. **6**(6), 248–254 (2002)

3. Baron-Cohen, S., Ashwin, E., Ashwin, C., Tavassoli, T., Chakrabarti, B.: Talent in autism: hyper-systemizing, hyper-attention to detail and sensory hypersensitivity. Philos. Trans. R. Soc. B: Biol. Sci. **364**(1522), 1377–1383 (2009)

4. Cabibihan, J.J., Pattofatto, S., Jomâa, M., Benallal, A., Carrozza, M.C., Dario, P.: The conformance test for robotic/prosthetic fingertip skins. In: The First IEEE/RAS-EMBS International Conference on Biomedical Robotics and Biomechatronics, BioRob 2006, pp. 561–566. IEEE (2006)

5. Cabibihan, J.J., Carrozza, M.C., Dario, P., Pattofatto, S., Jomaa, M., Benallal, A.: The uncanny valley and the search for human skin-like materials for a prosthetic fingertip. In: 2006 6th IEEE-RAS International Conference on Humanoid Robots, pp. 474–477. IEEE (2006)

6. Cabibihan, J.J., Javed, H., Sadasivuni, K., Al Haddad, A.: Smart robotic therapeutic learning toy, WIPO Patent WO2018033857, 22 Feb 2018 (2018)

7. Cabibihan, J.J., Javed, H., Aldosari, M., Frazier, T.W., Elbashir, H.: Sensing technologies for autism spectrum disorder screening and intervention. Sensors 17(1), 46 (2016)

8. Cabibihan, J.J., Javed, H., Ang Jr., M., Aljunied, S.M.: Why robots? A survey on the roles and benefits of social robots in the therapy of children with autism. Int. J. Soc. Robot. 5(4), 593–618 (2013)

9. Cabibihan, J.J., Joshi, D., Srinivasa, Y.M., Chan, M.A., Muruganantham, A.: Illusory sense of human touch from a warm and soft artificial hand. IEEE Trans. Neural Syst. Rehabil. Eng. 23(3), 517–527 (2015)

10. Cabibihan, J.J., So, W.C., Nazar, M., Ge, S.S.: Pointing gestures for a robot mediated communication interface. In: International Conference on Intelligent Robotics and Applications, pp. 67–77. Springer, Berlin, Heidelberg (2009)

11. Cabibihan, J.J., So, W.C., Pramanik, S.: Human-recognizable robotic gestures. IEEE Trans. Auton. Ment. Dev. 4(4), 305–314 (2012)

12. Cabibihan, J.J., So, W.C., Saj, S., Zhang, Z.: Telerobotic pointing gestures shape human spatial cognition. Int. J. Soc. Robot. 4(3), 263–272 (2012)

13. Eckerman, C.O., Whatley, J.L.: Toys and social interaction between infant peers. Child Dev. 48, 1645–1656 (1977)

14. Ferrari, E., Robins, B., Dautenhahn, K.: Therapeutic and educational objectives in robot assisted play for children with autism. In: The 18th IEEE International Symposium on Robot and Human Interactive Communication, RO-MAN 2009, pp. 108–114. IEEE (2009)

15. Golan, O., Ashwin, E., Granader, Y., McClintock, S., Day, K., Leggett, V., Baron-Cohen, S.: Enhancing emotion recognition in children with autism spectrum conditions: an intervention using animated vehicles with real emotional faces. J. Autism Dev. Disord. 40(3), 269–279 (2010)

16. Ham, J., Bokhorst, R., Cuijpers, R., van der Pol, D., Cabibihan, J.J.: Making robots persuasive: the influence of combining persuasive strategies (gazing and gestures) by a storytelling robot on its persuasive power. In: International Conference on Social Robotics, pp. 71–83. Springer, Heidelberg (2011)

17. Ham, J., van Esch, M., Limpens, Y., de Pee, J., Cabibihan, J.J., Ge, S.S.: The automaticity of social behavior towards robots: the influence of cognitive load on interpersonal distance to approachable versus less approachable robots. In: International Conference on Social Robotics, pp. 15–25. Springer, Heidelberg (2012)

18. Hart, M.: Autism/excel study. In: Proceedings of the 7th International ACM SIGACCESS Conference on Computers and Accessibility, pp. 136–141. ACM (2005)

19. Javed, H., Cabibihan, J.J., Al-Attiyah, A.A.: Autism in the Gulf states: why social robotics is the way forward. In: 2015 5th International Conference on Information & Communication Technology and Accessibility (ICTA), pp. 1–3. IEEE (2015)

20. Javed, H., Cabibihan, J.J., Aldosari, M., Al-Attiyah, A.: Culture as a driver for the design of social robots for autism spectrum disorder interventions in the middle east. In: International Conference on Social Robotics, pp. 591–599. Springer, Cham (2016)

21. Javed, H., Connor, O.B., Cabibihan, J.J.: Thomas and friends: implications for the design of social robots and their role as social story telling agents for children with autism. In: 2015 IEEE International Conference on Robotics and Biomimetics (ROBIO), pp. 1145–1150. IEEE (2015)

22. Lee, W., Cabibihan, J., Thakor, N.: Bio-mimetic strategies for tactile sensing. In: SENSORS 2013, pp. 1–4. IEEE (2013)

23. Machalicek, W., O'Reilly, M.F., Beretvas, N., Sigafoos, J., Lancioni, G.E.: A review of interventions to reduce challenging behavior in school settings for students with autism spectrum disorders. Res. Autism Spectr. Disord. 1(3), 229–246 (2007)

24. Michaud, F., Larouche, H., Larose, F., Salter, T., Duquette, A., Mercier, H., Lauria, M.: Mobile robots engaging children in learning. In: Canadian Medical and Biological Engineering Conference (2007)

25. Michaud, F., Duquette, A., Nadeau, I.: Characteristics of mobile robotic toys for children with pervasive developmental disorders. In: IEEE International Conference on Systems, Man and Cybernetics, vol. 3, pp. 2938–2943. IEEE (2003)

26. Parsons, S., Mitchell, P., Leonard, A.: The use and understanding of virtual environments by adolescents with autistic spectrum disorders. J. Autism Dev. Disord. 34(4), 449–466 (2004)

27. Peng, Y.H., Lin, C.W., Mayer, N.M., Wang, M.L.: Using a humanoid robot for music therapy with autistic children. In: Automatic Control Conference (CACS), 2014 CACS International, pp. 156–160. IEEE (2014)

28. Pop, C.A., Simut, R., Pintea, S., Saldien, J., Rusu, A., David, D., Vanderfaeillie, J., Lefeber, D., Vanderborght, B.: Can the social robot probo help children with autism to identify situation-based emotions? a series of single case experiments. Int. J. Humanoid Robot. 10(03), 1350025 (2013)

29. Ramdoss, S., Lang, R., Mulloy, A., Franco, J., O'Reilly, M., Didden, R., Lancioni, G.: Use of computer-based interventions to teach communication skills to children with autism spectrum disorders: a systematic review. J. Behav. Educ. 20(1), 55–76 (2011)

30. Robins, B., Dautenhahn, K., Te Boekhorst, R., Billard, A.: Effects of repeated exposure to a humanoid robot on children with autism. In: Designing a More Inclusive World, pp. 225–236. Springer (2004)

31. Robins, B., Otero, N., Ferrari, E., Dautenhahn, K.: Eliciting requirements for a robotic toy for children with autism-results from user panels. In: The 16th IEEE International Symposium on Robot and Human interactive Communication, RO-MAN 2007, pp. 101–106. IEEE (2007)

32. Sadasivuni, K., Al Haddad, A., Javed, H., Yoon, W., Cabibihan, J.J.: Strain, pressure, temperature, proximity, and tactile sensors from biopolymer composites. In: Biopolymer Composites in Electronics, pp. 437–457. Elsevier (2017)

33. Scassellati, B., Admoni, H., Mataric, M.: Robots for use in autism research. Ann. Rev. Biomed. Eng. 14, 275–294 (2012)

34. Shamsuddin, S., Yussof, H., Hanapiah, F.A., Mohamed, S., Jamil, N.F.F., Yunus, F.W.: Robot-assisted learning for communication-care in autism intervention. In: 2015 IEEE International Conference on Rehabilitation Robotics (ICORR), pp. 822–827. IEEE (2015)

35. Shamsuddin, S., Yussof, H., Ismail, L.I., Mohamed, S., Hanapiah, F.A., Zahari, N.I.: Initial response in HRI-a case study on evaluation of child with autism spectrum disorders interacting with a humanoid robot Nao. Procedia Eng. **41**, 1448–1455 (2012)
36. So, W.C., Wong, M.Y., Cabibihan, J.J., Lam, C.Y., Chan, R.Y., Qian, H.H.: Using robot animation to promote gestural skills in children with autism spectrum disorders. J. Comput. Assist. Learn. **32**(6), 632–646 (2016)
37. Teo, H.T., Cabibihan, J.J.: Toward soft, robust robots for children with autism spectrum disorders. In: FinE-R@ IROS, pp. 15–19 (2015)
38. Wykowska, A., Kajopoulos, J., Obando-Leitón, M., Chauhan, S.S., Cabibihan, J.J., Cheng, G.: Humans are well tuned to detecting agents among non-agents: examining the sensitivity of human perception to behavioral characteristics of intentional systems. Int. J. Soc. Robot. **7**(5), 767–781 (2015)
39. Zablotsky, B., Black, L.I., Maenner, M.J., Schieve, L.A., Blumberg, S.J.: Estimated prevalence of autism and other developmental disabilities following questionnaire changes in the 2014 national health interview survey. Natl. Health Stat. Rep. **87**, 1–21 (2015)
40. Zheng, Z., Young, E.M., Swanson, A.R., Weitlauf, A.S., Warren, Z.E., Sarkar, N.: Robot-mediated imitation skill training for children with autism. IEEE Trans. Neural Syst. Rehabil. Eng. **24**(6), 682–691 (2016)

iBridge - Participative Cross-Generational Approach with Educational Robotics

Georg Jäggle[1], Markus Vincze[1], Astrid Weiss[3],
Gottfried Koppensteiner[2], Wilfried Lepuschitz[2],
and Munir Merdan[2(✉)]

[1] ACIN Institute of Automation and Control, Vienna University of Technology,
Vienna, Austria
{jaeggle,vincze}@acin.tuwien.ac.at
[2] PRIA Practical Robotics Institute, Vienna, Austria
{koppensteiner,lepuschitz,merdan}@pria.at
[3] Institute of Visual Computing and Human-Centered Technology, Vienna
University of Technology, Vienna, Austria
astrid.weiss@tuwien.ac.at

Abstract. Robotics is both a technology and a factor that can be used to increase the interest of pupils and students in science and engineering. The project iBridge targets this issue and linked vocational schools and universities with older adults and students in order to increase their interest in innovative technology. This project is a cross-generational project that will engage the students in social and cross-cultural research topics through the application of robotics. In this paper, we introduce the project that allows students to take over the role of a co-researcher who investigate the demands and requirements of older adults towards an assistive service robot. The task of the pupils is to introduce the older adults to computer technology at a Pensioners' Club and develop a sensitive cuddly toy as a service robot within a series of workshops.

Keywords: Educational robotics · Participative research · Vocational school
Older adults · Sensitive cuddly toy

1 Introduction

Approximately 1.9 million people in Austria are over the age of 60; in total 23.5% of the population.[1] The proportion of people aged 60-plus will rise rapidly in the coming years, up to 30% by 2030, and as high as 33% by 2045.[2] Service robots are a way of enabling older adults to maintain independent living by supporting them in everyday life. Although this area promises secure jobs in the future, there is a lack of young researchers or students interested in pursuing robotics. The aim of this approach is to get children and young people involved in the field of assistive technologies for senior citizens. Moreover, it is of vital importance to address the needs and fears of older

---

[1] https://www.ffg.at/news/die-ersten-50-smart-homes-f-r-selbstbestimmtes-leben-im-alter.
[2] http://www.bifeb.at/fileadmin/user_upload/doc/Wissensgesellschaft._Bildungschancen_fuer_
aeltere_Menschen.pdf.

adults and understanding their desire to use new technologies and participate in new developments [1]. This starts with learning to use new information and communication technologies, which is often harder for older people due to diminished sensory capabilities (seeing and hearing) compared to the younger generation.

Nevertheless, although sufficient knowledge about problems with operation, handling, functionality, complexity, and corresponding improvement options has been available for many years, very few newly developed technologies and programs are age-appropriate, user-friendly designed, or barrier-free operable. The reasons for this are a certain degree of expert blindness and a lack of knowledge of the real needs and abilities of older people, as well as disregard for logical and natural procedures [3]. The study "My friend the robot" shows that service robotics certainly opens up positive potential for older adults: it allows them to facilitate their autonomous lifestyle, to support their health and safety at home, and to reduce dependency and the need for care [2]. Acceptance requirements should also become much more important in robot development than before [2]. For these reasons, qualitative examination methods seem sensible in order to analyze fears, worries, but also the positive expectations of the user groups who are referred to use service robotics [4]. The project iBridge faces the challenge of sensitizing the respective user groups to assistive technology usage, particularly socially assistive care robots, with the help of a sensitive cuddly toy prototype and including the potential end users in the development process. In addition, the development of the prototype serves to inspire children and young people. The project is based on cross-generational co-design workshops that include both young and old participants. In general, Austrian vocational schools are designed to teach the theory and practice of robotics applications, but students usually do not receive training in social and communication skills in addition to vocational theory and practice. Because of that, one of additional aim of the project is to explore how students acquire content knowledge[3] and social and communication skills[4] during a robotic research project.

The paper is structured as follows: The following Sect. 2 briefly introduces the project activities and their related schedule. Section 3 describes the different types of workshops, which are organized within the framework of the project. Section 4 gives a short introduction to the in-depth interviews pupils will perform with older adults in order to understand their requirements and needs towards an assistive robot. Section 5 presents the cuddly toy robot framework used within the project. Finally, the impact of the project is described in the Sect. 6 and a conclusion is given in the Sect. 7.

---

[3] https://www.tgm.ac.at/images/IT_Lehrplan.pdf.

[4] http://www.htl.at/fileadmin//content/Lehrplan/HTL_VO_262_2015/BGBl_II_Nr_262_2015_Anlage_1.pdf.

Fig. 1. Project schedule

2 Project Activities and Schedule

In the following we present the four concept phases (see Fig. 1) of the iBridge project and how they are integrated in the project process, including the underlying methodology:

Phase 1: In the first phase, the interactions of pupils with senior citizens are used to collect user wishes and requirements. After an analysis of the tasks and activities of older people, students will analyze the specific needs and demands of older people. In conversation with senior citizens, general needs and technical requirements are defined. A specification of the requirements finally takes place with the help of t observing hoe seniors are using existing computing technology. Additionally, research interviews are conducted with them during workshops (see Sect. 3). The resulting findings are compared with the results of the requirement analysis phase of the EU project Hobbit [12] and together form the basis for the evaluation and selection of the concepts and functions for the sensitive cuddly toy robot.

Phase 2: In the second phase, a first concept of the sensitive cuddly toy robot will be developed based on the results of the user survey, including the technical feasibility study. This concept serves as the basis for the first cycle of a dual-iterative design concept, which consists of prototyping and evaluation. A total of four groups of students will deal with different aspects of the development of the cuddly toy robot (see Sect. 5: Cuddly toy robot).

Phase 3: Proceeding from the concept, a first prototype is taken into operation and used for the first evaluation phase. During this phase, the prototype of the senior citizens will be tested to identify possible improvements to the hardware and software in the early phase of the project.

Phase 4: The fourth phase will be reviewed and refined based on the results of the first evaluation of the prototype. This updated concept serves as the starting point for the second cycle of the dual-iterative design concept, which consists of prototyping and evaluation. Within the technical implementation of the updated concept, the basic functions of the original concept are improved, and the identified comfort functions are also integrated into the current system while taking into account the first evaluation results. Based on the updated concept, a second prototype will be developed and put into operation, which will be used for the second evaluation phase. In the second evaluation phase, the updated prototype will be tested by seniors in realistic environments in order to identify the hardware and software improvements necessary in the advanced phase of the project. The gained insights from the evaluation should, among technical improvements, also be used in the creation of a user manual.

3   The Workshops and Cross-Generational Setting

The purpose of the workshops is to introduce students into robotics, but also to learn about the user requirements of older adults. Within the workshops at the TU Wien (Austria) students learn about existing robots for supporting older adults to get an idea of the technological capabilities of state-of-the-art systems. On the other side, within the framework of the workshops at the Practical Robotics Institute Austria, students learn about electrical, mechanical, and programming aspects of robotics. The purpose of the workshop is to enable students to fully conduct their own robotics projects. At the end, during the workshops in so-called Pensioners' Club (i.e. PC and Internet courses for older adults in retirement homes), students support older adults in using modern technologies within the context of PC and internet courses. The students profit from these courses by additionally interviewing the older adults and gaining knowledge about their needs, which are essential for successful service and assistive robotics solutions of the future.

(a)   Workshops at TU Wien

The workshops take place at the Automation and Control Institute (ACIN) at TU Wien. The design of the workshops is based on the didactic model of AVIVA, which is a model of classroom management and lesson planning in schools (see Fig. 2). This model links to the living world of the students and combines the instruction of knowledge with the application of knowledge. The last phase makes it possible to evaluate the different goals of the workshop. This alternation from instruction to construction is a perfect framework for a robotics workshop, because one of its goals is to pass on information about technical knowledge and robotics and to achieve a better result in acquisition of knowledge. Moreover, the students have a part to play in applying this knowledge. This model includes five different phases, although single parts can be multiply used, meaning that not all phases have to be performed. The different phases are called, in temporal sequence: "arrive, activate the previous knowledge, instruction, construction or application and evaluate" [10].

There are two main goals for the workshops. The first one is to show students the robotic technology and current robotic projects. The second one is to instruct them in research methodologies. The students will learn two qualitative methods: the half-structured research interview, and keeping a research diary [11].

The phase "Arrive"
In this phase, the students receive information about the schedule and their role in the workshop.

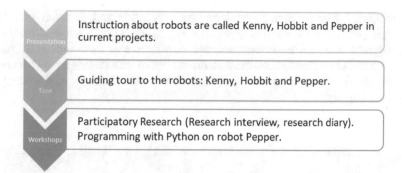

Presentation — Instruction about robots are called Kenny, Hobbit and Pepper in current projects.

Tour — Guiding tour to the robots: Kenny, Hobbit and Pepper.

Workshops — Participatory Research (Research interview, research diary). Programming with Python on robot Pepper.

Fig. 2. Workshop design

The phase "Activate the previous knowledge"
In this phase, the students first receive information about robots that is linked to their living world. There is a presentation combined with a discussion about robotics in their daily lives, robotics in the industry, and finally a discussion about service robotics at Automation and Control Institute (ACIN) at TU Wien.

The phase "Instruction"
This part of the workshop contains a presentation about the robots at the TU Wien given by a researcher, who explains the different components, properties, and demands on robots based on current projects. Subsequently, the students can see the different robots in reality and hear a detailed explanation of each robot (see Fig. 3).

The phase "Construction or application"
In this phase, the students work actively on different topics. One group has the option to learn about research interviews and research diaries during an instruction session about the rules and possibilities of research diaries, and subsequently to conduct

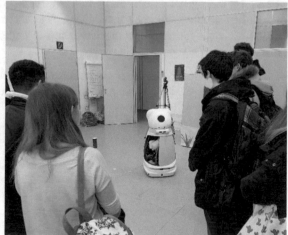

Fig. 3. Hobbit and Kenny

interviews with their partners to explore their perception and demands of robotics. During this part, the students gain experience regarding the problems and possibilities associated with research interviews and research diaries at their setting in the Pensioners' Club. The other group focuses on the robot called Pepper and learns the difference between the software "Choreographe" from Softbank and coding with Python (see Fig. 4).

Fig. 4. Programming and participative research

(b) Workshops at the Practical Robotics Institute Austria

The workshops at PRIA are intended to give students an effective introduction to robotics and the Hedgehog controller in a setting with an instructor teaching a limited number of grouped participants. The students are mentored in different ways of performing their tasks and how to solve the problems from stage to stage [13]. Throughout these activities, the student groups develop different ideas and concepts in order to accomplish their goals, issues, and problems. At the very beginning the students build up their first simple robot in electronic lessons to get in touch with an introductory level of robotics. In a further stage the workshop is dedicated to hands-on activities that include:

- Learning about robot motors, parts, sensors, and microcontrollers
- Learning how to design the mechanical structure of a robot
- Learning how to build, program and test robots. The repertoire includes graphical programming, but also easy programming guides for C-language. Different technologies will be used for the workshops such as LEGO Mindstorms or Botball[5] (see Fig. 5).

In the final stage, we provide students with the Hedgehog controller and help them build a robot they like. PRIA created the Hedgehog [14] with the intention of creating a suitable resource for anyone who wants to explore robotics, no matter their age or

---

[5] http://www.botball.org/

Fig. 5. Botball robot

expertise. Depending on the level of experience, the user can either employ the beginner-friendly, in-browser development environment, or use SSH to access the Hedgehog's Raspberry Pi directly.

The Hedgehog can be augmented using thousands of third-party applications & libraries for the Raspberry Pi, so the possibilities for this product are practically endless. It also supports both visual and textual programming through Blockly and Python. This stage encompasses the design of the robot based on the student ideas and experiences and focus on work in a team and the exchange of knowledge between them. It is a very practical stage with short lectures and a lot of time to try and discover. The workshop will be completed with a robot showcase.

(c) **The Trainer and Co-researcher in the Pensioners' Club**

This workshop takes place in the Pensioners' Clubs. The students introduce older adults to the technical basics of computers. They visit the older adults within a framework of six workshops to introduce them to using computers and to observe their needs and demands. Especially the young generation – the digital natives – profit from these courses by gaining knowledge about the problems that older adults have when using the PC and internet. Technology has predominantly been designed and developed as an organic medium for users who have grown up in the digital age, therefore it is of vital importance that students are learning the importance of recognizing the diverse attitudes and beliefs, dispositional characteristics, education and socio-economic status influence how older adults view and use Information and Communication Technologies (ICTs). Service and assistive robotics solutions of the future as well as ICT programs and technologies for older adults need to take into consideration this populations' characteristics, attitudes, and beliefs about ICTs, and in doing so they are more likely to accommodate their needs and interests [18]. During the last session, the students conduct a research interview, which is based on their observational data how older adults used the PC and the Internet and a half-structured interview (Sect. 4). The students change their role from trainer to Co-Researcher [8].

4   Research Design

The project is committed to inspiring adolescents to learn more about assistive technologies for senior citizens through access to robotics. This requires addressing the needs and fears of older adults and understanding their desire to use new technologies. [5] The adolescents will conduct research on the needs and fears of older people, and analyze the collected data from their own perspective in order to reduce the effects of expert blindness and a lack of knowledge of the real needs and abilities of older people [3]. Participative research includes use of research methodologies in which adolescents collect the data and conduct research to obtain a different view on the needs and abilities of older people.

Young people learn about the needs and demands of older people in connection with robotics and science. The development of robotics must focus on the living worlds of these people. Robotics is a technology of the future that has impressed young people in their professional field of activity. This project will promote a technical understanding of robotics and foster the participants' self-efficacy in relation to the curriculum of vocational schools (see footnote 4). This goal will be measured using the following research methods:

- In-depth interview
- Mind mapping
- Observation.

The in-depth interview is used to gain access to the seniors' social world through a half-structured conversation. The co-researcher obtains full descriptions and explanations from the older adults in their own words as much as possible [6, 7].

Mind mapping has been used in a variety of contexts and has developed into a tool generally used to represent an individual's knowledge and ideas about one particular theme [9]. The observational data is collected in research diaries from the students (see Fig. 6).

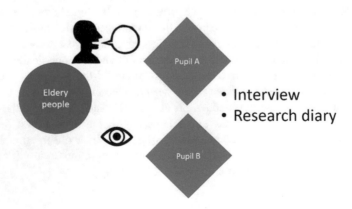

Fig. 6. Co-researcher setting

5 Sensitive Cuddly Toys

Studies have shown that animals have a capability to lower stress, reduce heart and respiratory rate, and show positive changes in hormonal levels, as well as mood elevation and social facilitation. Nevertheless, since they are not always compatible with certain lifestyles or easy to take care of and a new branch of robot pets has begun. The aim of robotic pet therapy is to create robots which can act as surrogates for those who do not have access to animals [15]. In the last years, there is also a growing number of robot toys such as Paro (robotic seal), Sony's AIBO (robotic dog), NeCoRo (robotic cat) and RIBA (bear robot), some of which have been mooted as possible companions for the elderly [16, 17].

Figure 7 shows current sensitive cuddly toy implementations, the Fruity Dave and Brum Baer, developed by the technical high school students within their diploma thesis. They developed robots in the form of a sensitive cuddly toy to be perceived as a cute and friendly interaction partner. The form of teddy toys was selected since it has a wide appeal among all age groups and are commonly found in the home. The toy consists of a series of body regions – the arms, the legs, the head, and body. The systems consists of a microcontroller (Raspberry Pi A+), tactile sensors (left arm and right arm inside), acceleration sensor, light sensors as well as a wireless module and a power bank. In the head of the robot are several multicolour LEDs and a speaker in the mouth. The cute sound effects sensitive cuddly toy produce are intended to elicit positive responses from users.

After increasing the interest of the young people and supporting the older generation towards modern technologies within the Co-Researcher in the Pensioners' Club workshops (Sect. 3), the co-design workshops are planned for creating the concept of an assistive robot as sensitive cuddly toy. As a basis for the development of the distinct parts of the concept, the requirements learned in the Penioners' Club, from the Hobbit" project [12], as well as other experiences [19, 20] are going to play a significant role. The aim for the students is to develop a robotic prototype that meets the needs of older people and improves their well-being.

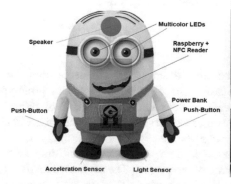

Fig. 7. Cuddly toys: Fruity Dave and Brum Baer

6 Impact

As an intergenerational project, which aims at increasing the interest of young people for research in cross social and cultural scenarios as well as to sensitize the older people for assistive technologies, the project iBridge is showing significant impact at an educational as well as on a social level. Three Vienna high schools Higher Technical School Wexstraße (TGM), Higher Technical School Donaustadt and Higher Technical School Ottakring together with researchers from PRIA and ACIN are jointly participating in the project.

Currently more than 150 students covering age group (15–19) participate in the project activities (see Fig. 8). They are working now in the first project stage visiting several retirement homes, which are also integrated in the project, for the Pensioners' Clubs.

Fig. 8. High school project conference at the TGM

7 Conclusion

This paper introduced iBridge, a project that aims at introducing students from vocational schools into the user-centered design process of at developing a robotic prototype which meets the actual requirements and needs of older adults. Through many different topics and entry-points (workshops with seniors, research and development in the fields of ICTs and mechanics), the participating students are given the opportunity to enter the scientific world and even finish for example their high school education with a corresponding diploma thesis within the frame of the project. The students take over an expert role in the research and technical part of the project. They learn to understand the technical opportunities robotics presents, in addition to the needs and demands of older adults, and research methods. By knowing that they contribute to an actual research project and solution of a relevant problem, the students are assured that their work is meaningful.

The future objective is to obtain results about the needs, demands, and perceptions of the older regarding robotics, what the students learned during the project, what the students' perceptions of robotics are, and how much influence the project had on the self-efficacy of the students.

Acknowledgements. The authors acknowledge the financial support by the Sparkling Science program, an initiative of the Austrian federal ministry of science and research, under grant agreement no. SPA 06/294.

References

1. ÖIAT: Studie: Maßnahmen für Senior/dinner in der digitalen Welt. Wien: BMASK (2014). www.saferinternet.at/senioren-digitale-welt
2. Meyer, S.: Mein Freund der Roboter: Servicerobotik für ältere Menschen – eine Antwort auf den demographischen Wandel?. VDE Verlag, Berlin (2011)
3. Mollenkopf, H.: Chancen und Barrieren der neuen Medien – Technik auch im Alter kompetent nutzen am 10. Enquete Alter hat Zukunft - gerne älter werden in Tirol Internet verbindet – Fit fürs Informationszeitalter, Innsbruck (2006)
4. Krings, B.-J., Böhle, K., Decker, M., et al.: ITA-Monitoring Serviceroboter in Pflegearrangements. Kurzstudie. Karlsruhe: Erschienen. In: Decker, M., Fleischer, T., Schippl, J., Weinberger, N. (Hrsg.) Zukünftig Themen der Innovations - und Technikanalyse. Lessons Learned und ausgewählte Ergebnisse. KIT Scientific Reports 7668. Karlsruhe: KIT Scientific Publishing, S. 63–121 (2014)
5. Glende, S.: Senior User Integration - Konzepte, Werkzeuge und Fallbeispiele: Ein Leitfaden für die Einbindung älterer Nutzer in die Produkt- und Serviceentwicklung. Saarbrücken (2010)
6. Kelly, A.P.: Social research methods, p. 53. University of London (2016)
7. Neumann, W.: Social Research Methods: Qualitative and Quantitative Approaches, p. 461. Pearson, London (2014)
8. Smith, R., Monaghan, M.: Involving young people as co-researchers: facing up to the methodological issues. Qual. Soc. Work **1**(2), 191–207 (2015)
9. Meier, P.S.: Mind-mapping: a tool for eliciting and representing knowledge held by diverse informants. Soc. Res. Update **52**, 1–4 (2007)
10. Städeli/Grassi/Rhiner/Obrist: Kompetenzorientiert Unterrichten. Das AVIVA-Modell, hep Verlag (2010)
11. McTaggart, R.: Participatory Action Research: International Contexts and Consequences. Suny Press, Albany (1997)
12. Vincze, M., Weiss, A., Lammer, L., Huber, A., Gatterer, G.: On the discrepancy between present service robots and older persons' needs. In: 23rd IEEE International Symposium on Robot and Human Interactive Communication (IEEE RO-MAN 2014), 25–29 August 2014, Edinburgh (2015). http://hobbit.acin.tuwien.ac.at/publications/hobbit_roman.pdf. Accessed 30 Jan 17

13. Lepuschitz, W., Koppensteiner, G., Merdan, M.: Offering multiple entry-points into STEM for young people. In: Robotics in Education - Research and Practices for Robotics in STEM Education, Advances in Intelligent Systems and Computing, pp. 41–52. Springer (2017). ISBN 978-3-319-42974-8

14. Krofitsch, C., Hinger, C., Merdan, M., Koppensteiner, G.: Smartphone driven control of robots for education and research. In: IEEE 2013 International Conference on Robotics, Biomimetics, Intelligent Computational Systems, Yogyakarta, Indonesien (2013)

15. Stiehl, W.D., Lieberman, J., et al.: Design of a therapeutic robotic companion for relational, affective touch. In: Proceedings of the IEEE International Workshop on Robot and Human Interactive Communication (ROMAN 2005), pp. 408–415 (2005)

16. Sharkey, A., Sharkey, N.: Granny and the robots: ethical issues in robot care for the elderly. Ethics Inf. Technol. 14(1), 27–40 (2012)

17. Lazar, A., Thompson, H.J., Piper, A.M., Demiris, G.: Rethinking the design of robotic pets for older adults. In: Proceedings of the 2016 ACM Conference on Designing Interactive Systems, pp. 1034–1046 (2016)

18. Vroman, K.G., Arthanat, S., Lysack, C.: Who over 65 is online?" Older adults' disposition toward information communication technology. Comput. Hum. Behav. 43, 156–166 (2015)

19. Payr, S., Werner, F.: Potential of Robotics for Ambient Assisted Living. Final Report, benefit/BMVIT (2015)

20. Liu, L., Stroulia, E., Nikolaidis, I., Miguel-Cruz, A., Rios Rincon, A.: Smart homes and home health monitoring technologies for older adults: a systematic review. Int. J. Med. Inform. 91, 44–59 (2016)

Modelling the Driver Assistance Systems Using an Arduino Compatible Robot

Richard Balogh[✉] and Peter Ťapák

Slovak University of Technology in Bratislava, Bratislava, Slovakia
{richard.balogh,peter.tapak}@stuba.sk

Abstract. It is difficult to demonstrate functions of Advanced Driver Assistance Systems in the classroom – they are usually proprietary and too complex, expensive and non-repetitive. In the paper we describe some functions of the driver assistance systems and their operation. Then we show the possibility of their modeling using a mobile robot. We used the Acrob, Arduino compatible robot, as a model robot. All the experiments proved that it is feasible and beneficial approach applicable very well into the curricula.

Keywords: Acceleration skid control · Track control system
Autonomous cruise control · Active lane assistant
Advanced driver assistance systems

1 Introduction

On the road to the fully autonomous car there are two approaches. First, to build the whole car from the scratch as proposed by the information technology companies (see e.g. Google car[1]). On the other side, car producing companies with more experiences from the automotive industry prefer step-by-step approach, improving existing safety and driver assistance functions (see e.g. Volkswagen, Audi and other cars with more or less semi-autonomous functions), which often come from third parties.

When teaching automotive mechatronic systems, the driver assistance functions are important part of curricula. Unfortunately, the real examples are hard to deliver, sometimes even impossible to test in real conditions (non-repetitive properties of an airbag activation system as an example). Teaching just theories, describing the function and properties of such system, even combining with animations and videos is also not the answer. So we tried to model some functions using the mobile robots, giving the students an opportunity to modify and test the functions in real systems. In the following paper we will provide three examples of such modeling, showing the possibilities of using robots outside the robotics curriculum.

---

[1] https://www.google.com/selfdrivingcar/.

© Springer Nature Switzerland AG 2019
W. Lepuschitz et al. (Eds.): RiE 2018, AISC 829, pp. 275–283, 2019.
https://doi.org/10.1007/978-3-319-97085-1_27

2   Driver Assistance Systems

Car safety and more generally the road safety have been in the focus of technological companies for a long time. Every year more than 1.17 million people die
in road accidents around the world [1]. Especially those alarming numbers of the
car accidents victims raises many initiatives from technological to political. EU
Commissioner for Transport Violeta Bulc said *"Every death or serious injury
is one too many. We have achieved impressive results in reducing road fatalities
over the last decades but the current stagnation is alarming. If Europe is to reach
its objective of halving road fatalities by 2020, much more needs to be done."* [2].
Besides the other precautions (e.g. identifying the safety hazardous locations,
modifying the travel speeds, etc.) the technology solution inside the car itself
is one of the main contributions to the overall road safety. Both the European
Union and the United States are mandating that all vehicles be equipped with
autonomous emergency-braking systems and forward-collision warning systems
by 2020 [3].

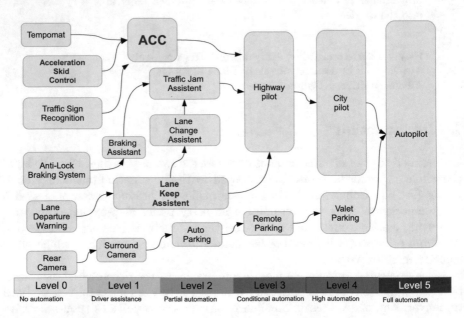

Fig. 1. Levels of automation on the road towards the fully autonomous car [4].

Advanced Driver Assistance Systems (ADAS) allow a vehicle to monitor near
and far fields in every direction using many various sensors and using sophisticated algorithms which ensure vehicle and personal safety based on detection of
changing factors such as weather conditions, trajectory and traffic conditions,
and detecting potential dangerous situations. Depending on the safety system
level in particular vehicle, the modern ADAS systems can act via warnings to

the driver or even by real actuation of the vehicle systems. ADAS systems can be considered as precursors to the fully autonomous vehicles of the future. Good overview of the driver assistance systems with animations and general information can be found online[2].

Yet in the early stage of development, there are many challenges to design reliable and robust ADAS. The system is expected to obtain reliable and accurate information from its sensors, be fast enough in processing those data and react in real time. To achieve low error rates, significant amount of effort and research has to be done. Moreover, in latest years the question of the security of such systems appears in the focus of developers and researchers [5].

3   Acceleration Skid Control (ASR)

Also known as a traction control system (TCS), is typically (but not necessarily) a secondary function of the electronic stability control (ESC) on production motor vehicles, designed to prevent loss of traction of driven road wheels. TCS is activated when throttle input and engine torque are mismatched to road surface conditions. The acronym ASR came from german *Antriebsschlupfregelung* – engine slippage regulation.

Since the 1971, when the Buick division of GM introduced MaxTrac as the first TCS, more and more sophisticated systems have been designed (e.g. Robert Bosch GmbH). Systems usually involve an engine management controller that cooperates with the brake system in order to prevent the driven wheels from spinning out.

System is an extension to an existing ABS (Anti-lock Braking System) system and cannot be installed alone. When the electronic control unit (ECU) recognizes wheel spinout by detecting sharp increases in wheel acceleration, then it reduces the engine torque through the engine management controller in a closed-loop process to reduce the traction force on the driven wheels [6].

More details about the controller and classical implementation can be found e.g. in [7]. Fuzzy controller is also a solution often used [8].

There are two ways in which ASR prevents the drive wheels from spinning. First of all it minimizes wheel spinning using metered brake force. Secondly it also regulates the engine torque via the electronic gas pedal. Even at full throttle the engine provides only as much output as the drive wheels can transfer. It is a big advantage for safety when starting off and for handling.

When starting off, the torque and the engine speed affecting the drive wheels are monitored. The torque distribution is controlled in such a way that wheel spinning is prevented, and an optimum flow of power is constantly guaranteed.

Generally, in car with ASR, each wheel is equipped with a speed sensor (see e.g. [9]). The ASR controls the slip of drive wheels, the wheels propelling the car. The slip is relative motion between a tire and the road surface it is moving on. Usually, wheel slip is calculated as a drive wheel speed relative to a speed

---

of non-driven wheels. If the drive wheel slip is over a threshold it means lost traction of this wheel and the control unit reduces the engine torque or brakes the slipping wheel. This applies to rear wheel drive or front wheel drive cars. For all wheel drive cars it is little more complicated to determine the speed of the car, but still the ASR controls the wheel slip of all driven wheels to be in specified limits.

Realization: One of the laboratory exercises of the lectures on Automotive movement processes was on the ASR. The Acrob mobile robot was used and programmed by the students in the C language. Unlike common cars the Acrob is a differential drive vehicle, therefore the traction control algorithm had to be changed to suit the capabilities of its sensors and actuators. Since there is no speed sensor on the only non-driven wheel, it is difficult to determine the wheel slip. Therefore the following algorithm is not wheel slip based, but only constrains the wheel acceleration to a safe limit in a closed loop.

Speed of both drive wheels is measured using an reflexive photocoupler connected to the corresponding counters inside the microcontroller. Then, using an interrupts, the actual speeds $v_L(k)$ and $v_R(k)$ in the moment k are determined. Calculating the difference we obtain an acceleration $a(k) = v_i(k) - v_i(k-1)/\Delta t$.

Its value is compared with the safe value a_S. Then the necessary adjustment is done on each wheel separately. Actual excerpt of the Arduino code follows:

```
if (iL)                 // if the interrupt from photocoupler
{
  iL = LOW;             // reset interrupt flag
  tLnew = millis();     // new timestamp
  // following    calculation is based on the wheel diameter
      and the number of interrupts per rotation
  vLnew = 0.026/((double)(tLnew-tLold)/1000.0); //speed [m/s]
  // wheel acceleration [m/s^2]
  aL = (vLnew-vLold)/((double)(tLnew-tLold)/1000.0);

  if (aL>a_S)
    {
      POWER=power_adjust();
    }
  LeftServo.writeMicroseconds(POWER);
  vLold = vLnew;        //store old timestamp
  tLold = tLnew;        //store old speed for next calculation
}
```

4   Autonomous Cruise Controll (ACC)

The cruise control (in some countries known as tempomat or speed control) is a system that automatically controls the speed of a road vehicle. The principle was first used in the Chrysler 1958 Imperial, based on the 1948 invention of a mechanical engineer Ralph Teetor. This system calculated ground speed based on driveshaft rotations, and used a bidirectional screw-drive electric motor to vary throttle position [1]. Adaptive (sometimes autonomous) cruise control (ACC) is an advanced cruise control system that automatically adjusts the vehicle speed to maintain a safe distance from vehicles ahead. Control is based on an information from on-board sensors. First commercial system in use was a lidar-based distance detection system in Mitsubishi Debonair in 1992. This system only warned the driver, without influencing throttle, brakes or gearshifting. The first commercial radar-assisted ACC system was used in 1999 by Mercedes-Benz in S-Class W220 model [2]. The control logic of the cruise controller can be designed by employing different types of controllers, such as a proportional-integral-derivative (PID) controller [3,4] or even with feedforward system only [8].

The cruise control maintains a constant vehicle speed, despite having external disturbance as a road inclination and wind speed or direction changes. The speed of the vehicle is measured (e.g. on the driveshaft) and compared to desired one. The controller sets the desired pedal position. We assume that we have an information on actual vehicle speed and relative distance information from the previous vehicle (this information interests us only if the distance is less than the preset distance). Instead of the safe distance it is sometimes preferable to use the safe time. i.e. the time at which we would approach a given vehicle at a given speed.

More information about modeling and design of the controller for the ACC system can be found in [13].

The following example was made at a laboratory exercise in subject autonomous mechatronic systems. The ultrasonic distance measurement sensor HC-SR04 was placed at the front of the Acrob robot. Since the servos sampling time is 20 ms the distance measurement was made with the same sampling time. The Arduino's timer 2 was used to generate the interrupt every 20 ms for this purpose. Moving average filter was applied to a measured data. Since the servos propelling the wheels have the speed controller already, there is just a proportional controller in the ACC loop, which provides setpoints to wheel speed control loops of the servos.

```
if (distance >= 400 || distance <= 0){
  Serial.println("Out of range");
}
else {
FIR[j]=distance; j++; j=j%10;    // moving average (FIR)
    filter
double Fdistance=0;
for(int i=0;i<10;i++)
{
   Fdistance+=FIR[i];
}
Fdistance=Fdistance/10;

u = Kp*(r-Fdistance);            // P - controller
u = constrain(u,-200,200);
```

5   Lane Departure Warning (LDW)

The Lane Departure Warning, known also as a Lane Assist (SPA) function uses a camera system mounted behind the windscreen to issue a warning if the vehicle is in danger of coming off the road. LDW continuously monitors the distance of the bus to the lane markings. As soon as the vehicle crosses the lane markings, the driver is warned through the corresponding side of the seat starting to pulsate. LDW is activated at speeds upwards of 70 km/h and is deactivated if the indicator is used, to indicate that the driver wishes to change lane, for example.

A lane detection system used behind the lane departure warning system can use the principle of Hough transform and Canny edge detector to detect lane lines from realtime camera images fed from the front-end camera of the automobile. A basic flowchart of how a lane detection algorithm works is shown in Fig. 2.

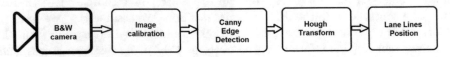

Fig. 2. Lane detection algorithm.

The image processing was not the topic of the lectures but we want to show the principle of operation. Inspiration came from the NxP Cup[3] – a worldwide competition for university teams to program an autonomous car for racing in loops marked with two black lines (see Fig. 3). In this competition, the cars use very simple single line camera to detect the lines [12].

Using a very simple visualization, based on a Processing script[4] enables students to view the signal in real time and within the real lighting conditions. As the time on the lab is usually short, we provide them with the code template, so they have enough time to play with the constants in the program to improve the image quality. Their task is then to design the appropriate algorithm for lane detection. Then they try to program the robot to maintain its position within the lanes.

```
/* part of the Arduino code for reading all 128 pixels into
   the array Pixel[] */

   for (i = 0; i < NPIXELS; i++) {
     Pixel[i] = analogRead (AOpin)/4 ;  // 8-bit is enough for
         our purpose
     digitalWrite (CLKpin, LOW);        // Each clock pulse
         shifts the next
     delayMicroseconds (1);             //     pixel value onto
         its output
     digitalWrite (CLKpin, HIGH);
   }
```

---

[3] See e.g. https://community.nxp.com/docs/DOC-1284.
[4] Full source http://senzor.robotika.sk/sensorwiki/index.php/TSL1401_Line_Sensor.

Fig. 3. Competition NxP Cup - following the trajectory marked with two lines.

As an example, sampled input signal from the camera is in the Fig. 4. Unfortunately, the camera optics is not perfectly suited for this sensor, so the edges of the signal (pixels 1–17 and 120–128) are too noisy to be useful. To help students with the visualization of what camera sensor sees, we have a simple Processing[5] application to display not only the current line, but also 127 previous ones so we obtain something like the photo of the lanes (see the top right in Fig. 4). It is then easy to detect the edges, find their center and move the vehicle so it maintains its position within the lanes and moving forward simultaneously. As an example, we entered also the maximum (red), minimum (green) and detected lane (yellow) pixels into the image.

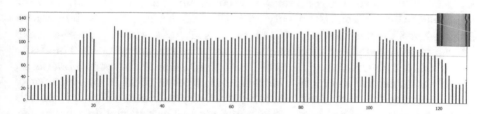

Fig. 4. Sampled signal from the camera is the array of 128 values. Simple processing application displays it together with its history.

---

[5] https://processing.org/.

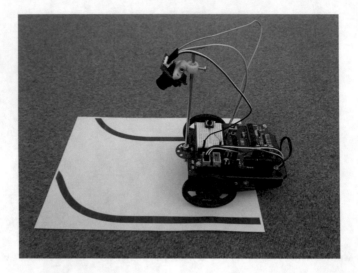

Fig. 5. Mobile robot Acrob with the line camera during the test.

6   Conclusions

In this paper we presented three models of the ADAS functions demonstrated using a small differential driven mobile robot Acrob. All of them were used during the previous summer semester in various lectures as a motivating tool to deepen the students topics interest. Students were in their 3rd year of bachelors degree studies or the 1st year of masters degree studies at the Institute of Automotive Mechatronics at the Slovak University of Technology in Bratislava. As their previous programming experiences were in many cases not sufficient enough and the time for these exercises was always limited to 90 min only, we usually provide the students with the program templates containing the core functions and the basic algorithm, leaving them only with the task to improve the function or program just the partial section of the program. All of the models presented can be also recommended as a more advanced long-term student projects. The Acrob is not like conventional car, it has different propulsion and dynamics, however, it is electric and the automotive industry puts a lot of effort in development of electric cars, so the most of the automotive industry engineers will have to be able to deal with conventional cars, hybrids and electric cars in the near future. As compared with conventional vehicles, the propulsion systems of electric vehicles can be realized with various powertrain options which differ in terms of the number of driving axles and wheels, arrangement and number of electric motors (see e.g. [9]). From the engineering point of view, individually controlled electric motors allow reduction of conventional friction brakes and higher bandwidth in control, all of which are provided by the proposed robot.

Acknowledgements. This paper was supported by the grants of the Research Agency of the Ministry of Education, Science, Research and Sport of the Slovak Republic (VEGA) 1/0819/17 Intelligent mechatronics systems (IMSYS) and its Cultural and Educational Grant Agency (KEGA) Kega 025STU-4/2017.

References

1. Mehar, R., Agarwal, P.K.: A systematic approach for formulation of a road safety improvement program in India. Procedia - Soc. Behav. Sci. **104**, 1038–1047 (2013). https://doi.org/10.1016/j.sbspro.2013.11.199. ISSN 1877-0428
2. European Commission - Press release: road safety – new statistics call for fresh efforts to save lives on EU roads, Brussels, 31 March 2016. http://europa.eu/rapid/press-release_IP-16-863_en.htm
3. Choi, S., Thalmayr, F., Wee, D., Weig, F.: Advanced driver-assistance systems: challenges and opportunities ahead. McKinsey Quarterly (2016). https://www.mckinsey.com/industries/semiconductors/our-insights/advanced-driver-assistance-systems-challenges-and-opportunities-ahead
4. J3016 201401: Taxonomy and Definitions for Terms Related to On-Road Motor Vehicle Automated Driving Systems. On-Road Automated Driving (ORAD) Committee, SAE International, 12 p. (2014). https://doi.org/10.4271/J3016_201401
5. Zha, M.: Advanced Driver Assistant System Threats, Requirements, Security Solutions. Technical White Paper, Intel Labs (2015). https://www.intel.com/content/dam/www/public/us/en/documents/white-papers/advanced-driver-assistant-system-paper.pdf
6. Jalali, K.: Stability control of electric vehicles with in-wheel motors. PhD. theses, University of Waterloo (2010)
7. Hori, Y., Toyoda, Y., Tsuroka, Y.: Traction control of electric vehicle: basic experimental results using the test EV UOT Electric March. IEEE Trans. Ind. Appl. **34**(5), 1131–1138 (1998)
8. Bauer, M., Tomizuka, M.: Fuzzy logic traction controllers and their effect on longitudinal vehicle platoon systems. Veh. Syst. Dyn. **25**(4), 277–303 (1996)
9. Ivanov, V., Savitski, D., Shyrokau, B.: A survey of traction control and antilock braking systems of full electric vehicles with individually controlled electric motors. IEEE Trans. Veh. Technol. **64**(9), 3878–3896 (2015). https://doi.org/10.1109/TVT.2014.2361860. NOT OK
10. Balogh, R.: Educational robotic platform based on Arduino. In: Proceedings of the 1st international conference on Robotics in Education, RiE2010, pp. 119-122 (2010)
11. Balogh, R: Robotics course with the Acrob robot. In: 2nd International Conference on Robotics in Education, pp. 41-46 (2011)
12. Fuentes, F.R., Trujillo, M., Padilla, C., Mendoza, R.: Using Parallax TSL1401-DB Linescan Camera Module for line detection. Freescale Semiconductor Application Note. Document Number: MPC5604B (2011). https://cache.freescale.com/files/microcontrollers/doc/app_note/AN4244.pdf
13. Chamraz, S., Balogh, R.: Two approaches to the adaptive cruise control (ACC) design. In: Proceedings of the 29th Cybernetics and Informatics (K&I) Conference. IEEE (2018)
14. Van Zanten, A.T.: Evolution of electronic control systems for improving the vehicle dynamic behavior. In: Proceedings of the 6th International Symposium on Advanced Vehicle Control, p. 9 (2002). http://citeseerx.ist.psu.edu/viewdoc/download?doi=10.1.1.582.6
15. Balogh, R., László, M.: Model-based design of a competition car. In: Robotics in Education. Advances in Intelligent Systems and Computing, vol. 457. Springer, Cham (2017). https://doi.org/10.1007/978-3-319-42975-5_20. ISBN 978-3-319-42974-8

Author Index

© Springer Nature Switzerland AG 2019
W. Lepuschitz et al. (Eds.): RiE 2018, AISC 829, pp. 285–286, 2019.
https://doi.org/10.1007/978-3-319-97085-1

Printed in the United States
By Bookmasters